高等学校土建类学科专业“十四五”系列教材
高等职业教育活页式教材

建筑英语口语

Oral English for Civil Construction

主　编　耿玉丽　王　珊　付万俊
副主编　邵　军　孟子寒　高　艳　王一兰
主　审　张克峰　宁雪峰

中国建筑工业出版社

图书在版编目（CIP）数据

建筑英语口语 = Oral English for Civil Construction / 耿玉丽，王珊，付万俊主编；邵军等副主编 . -- 北京：中国建筑工业出版社，2024. 8.（高等学校土建类学科专业“十四五”系列教材）（高等职业教育活页式教材）. -- ISBN 978-7-112-30020-4

Ⅰ . TU

中国国家版本馆 CIP 数据核字第 2024VJ4261 号

责任编辑：勾淑婷　王美玲
责任校对：李美娜

高等学校土建类学科专业“十四五”系列教材
高等职业教育活页式教材
建筑英语口语
Oral English for Civil Construction
主　编　耿玉丽　王　珊　付万俊
副主编　邵　军　孟子寒　高　艳　王一兰
主　审　张克峰　宁雪峰
*
中国建筑工业出版社出版、发行（北京海淀三里河路 9 号）
各地新华书店、建筑书店经销
北京雅盈中佳图文设计公司制版
北京市密东印刷有限公司印刷
*
开本：787 毫米 ×1092 毫米　1/16　印张：$11^1/_4$　字数：156 千字
2024 年 8 月第一版　2024 年 8 月第一次印刷
定价：**50.00** 元
ISBN 978-7-112-30020-4
（42885）

编写委员会

主　编：耿玉丽（山东商务职业学院）

　　　　王　珊（山东建筑大学）

　　　　付万俊（淄博市鲁中勘察设计审查咨询中心）

副主编：邵　军（山东商务职业学院）

　　　　孟子寒（山东智通慧达物流有限公司）

　　　　高　艳［威而德（日照）园林机械有限公司］

　　　　王一兰（山东商务职业学院）

主　审：张克峰（山东建筑大学）

　　　　宁雪峰（山东百伟建设工程管理有限公司）

编　委：朱　峰（山东商务职业学院）

　　　　王子晗（山东商务职业学院）

　　　　李海强（深圳市新城市规划建筑设计股份有限公司山东分公司）

　　　　唐卫东［中节能（山东）节能环保科技有限公司］

　　　　时信珂（山东蓝科工程咨询有限公司）

　　　　宋吉鑫（烟台飞龙集团有限公司）

Summary

Based on the construction engineering profession, this textbook uses a mechanical equipment factory as a construction project case to carry out project-based teaching in modules including the project decision-making, project preparation, preliminary design, construction drawing design, construction phase, completion phase, and a special section on Green Buildings. It presents the work processes and tasks required in each phase of the project in the form of oral situational dialogues. It enables learners to master various professional techniques in the project construction process, to use professional English correctly in dialogue, and to demonstrate their technical knowledge and opinions in a conversational manner.

This textbook is applicable to teaching and learning in higher vocational colleges and universities in the field of construction. It can also serve as a reference for engineering and technical personnel.

前言

大道致远，海纳百川。随着“一带一路”的推进和建设发展，当今世界合作更深入，距离不再遥远。倡议源于中国，但机会和成果属于世界，其旨在为“一带一路”建设指明方向，彰显中国担当和世界情怀。

党的二十大报告指出：推进高水平对外开放，稳步扩大规则、规制、管理、标准等制度型开放，加快建设贸易强国，推动共建“一带一路”高质量发展，维护多元稳定的国际经济格局和经贸关系。

建筑行业面临的是一个经济全球化、信息国际化、知识无界化、学习全民化、教育普及化的崭新时代。在“一带一路”思想的指导下，建筑行业展现出全新的形态——走出国门、助力世界。让语言沟通无障碍、让技术交流无国界是本教材编写的目的。

教材编写内容是将一个项目从决策阶段到竣工阶段的整个过程通过一些片段贯穿起来，让学生了解一个完整项目建设的始末流程和相关建筑专业知识，共包括 20 个模块学习任务。在使用过程中，你可以使用微信扫描书中的二维码展开学习。根据自身学习需求选择相关的学习内容。通过逐个任务的学习，你将逐渐提升口语流利度和应用能力。

本教材中，耿玉丽（工程技术应用研究员 / 山东商务职业学院）负责整部教材编写工作。王珊（项目副研究员 / 山东建筑大学）、付万俊（高级工程师 / 淄博市鲁中勘察设计审查咨询中心）参与部分专业知识编写。张克峰（二级教授 / 山东建筑大学）、宁雪峰（高级工程师 / 山东百伟建设工程管理有限公司）负责教材审核。邵军（山东商务职业学院）、王一兰（山东商务职业学院），承担整个教材语言文字校核与语法等修正工作。孟子寒（山东智通慧达物流有限公司）负责绘制全文插图。高艳[威而德（日照）园林机械有限公司建设项目经理及采购经理]负责英语口语化审核与语法修正工作。

在编写过程中，本书得到了淄博市鲁中勘察设计审查咨询中心、山东智通慧达物流有限公司、威而德（日照）园林机械有限公司等公司的大力支持，他们提出许多非常宝贵的建议和修改意见。本书也参考了诸多同类教材、文献，在此，向所有提供帮助的参与人员表示深深的感谢。

由于编者水平有限，教材中可能存在不妥之处，恳请广大读者批评指正。

编者

Perface

The great road leads to the distant horizon, and the ocean accommodates all rivers. With the advancement and development of "the Belt and Road" initiative, the world today has embraced cooperation and distances are no longer insurmountable. Although the initiative originated from China, the opportunities and achievements belong to the world. It aims to provide guidance for the construction of "the Belt and Road", and to demonstrate China's responsibility and global vision.

Report to the 20th National Congress of the Communist Party of China pointed out the need to promote high-level opening-up to the outside world, to steadily expand institutional openness in areas such as rules, regulations, management, and standards, to accelerate the construction of a strong trading nation, to promote high-quality development of jointly building "the Belt and Road", and to maintain a diverse and stable international economic order and trade relations.

The construction industry is facing a brand-new era of economic globalization, information internationalization, boundary-less knowledge, universal learning, and widespread education. Under the guidance of "the Belt and Road" concept, the construction industry has shown a new form—going global and contributing to the world. The purpose of this textbook is to promote barrier-free language communication and unrestricted technical exchange.

The content of the textbook covers various stages from the project decision-making to completion, by selected fragments, aiming to help students understand the complete process of project construction and relevant civil construction knowledge. It includes a total of 20 module learning tasks. You can use WeChat to scan the QR code in the book for learning. Choose relevant learning content referring to your learning demand. You will gradually improve your oral fluency and application skills through task learning.

Geng Yuli (Engineering Technology Application Researcher/ Shandong Business Institute) is responsible for the overall writing. Wang Shan (Project Associate Researcher/Shandong Jianzhu University) and Fu Wanjun (Senior Engineer/Luzhong Examing Consultative Center of Prospecting Designs Zibo City) participated in the writing of certain professional knowledge. Zhang Kefeng (Second Grade Professor/Shandong Jianzhu University) and Ning Xuefeng (Senior Engineer/Shandong Baiwei Construction Engineering Management Co., Ltd.) are responsible for textbook review. Shao Jun and Wang Yilan (Shandong Business Institute) are responsible for the language verification and grammar correction of the entire textbook.Meng Zihan (Shandong Zhitong Huida Logistics Co., Ltd.) is responsible for drawing text illustrations.Gao Yan (Construction Project Manager and Procurement Manager/Wild Rizhao Garden Machinery Co., Ltd) is responsible for English colloquialism review and grammar correction.

During the writing process, we have received strong support from Luzhong Examing Consultative Center of Prospecting Designs Zibo City, Shandong Zhitong Huida Logistics Co., Ltd., Wild Rizhao Garden Machinery Co., Ltd, and other companies. They put forward many valuable suggestions and comments. We have also referred to many similar textbooks and other literature. We would like to express our deep gratitude to all participants who have provided assistance.

Due to the limited skills of writers and editors, there may be shortcomings in the textbook. We kindly request readers to provide criticisms and corrections.

Writers and Editors

Directory

Diana—Investor	建设单位负责人
Alex—Vice General Manager	建设单位副总经理
Shrek—Project Manager	建设单位项目经理
Vivian—Plant Manager	新建工厂经理
Titi—Civil Engineer	建设单位土建工程师
Bard—Electrical Engineer	建设单位电气工程师
Mike—Equipment Engineer	建设单位设备工程师
Jodie—Design Manager	设计单位经理
Holger—Civil Engineer	设计单位土建工程师
Divid—Electrical Engineer	设计单位电气工程师
Sam—Equipment Engineer	设计单位设备工程师
Alvin—Architectural Designer	设计单位建筑设计师
Olay—Architectural Designer	设计单位建筑设计师
Peter—Architectural Designer	设计单位建筑设计师
Bruce—Structural Designer	设计单位结构设计师
Nate—Structural Designer	设计单位结构设计师
Leona—Structural Designer	设计单位结构设计师
Andy—Water Designer	设计单位给水排水设计师

Eric—Water Designer	设计单位给水排水设计师
Jacy—Heating and Air Conditioning Designer	设计单位供暖与空调设计师
Pope—Heating and Air Conditioning Designer	设计单位供暖与空调设计师
Amy—Electrical Designer	设计单位电气设计师
Cute—Electrical Designer	设计单位电气设计师
Divian—Tenderer	投标商
Chris—Tenderer	投标商
Vivid—Tenderer	投标商
Tallmi—Manager of Project Management Company	项目管理公司经理
Helen—Construction Project Manager	施工单位项目经理
Kevin—Civil Engineer	施工单位土建工程师
Mary—Equipment Engineer	施工单位设备工程师
Bob—Supervising Engineer	监理工程师
Jony—Piping Worker	管道安装工人
Kathy—Teacher	老师
Polly—Student	学生
Teddy—Student	学生
Michael—Student	学生
Linda—Student	学生

Module One

Project Decision Phase

Task 1　Task Description

Task 2　Oral Practice

Task 3　Analysis of Civil Construction Knowledge

Task 1　Task Description

Part Ⅰ　Task Description

Understand the procedures and content of construction projects during the decision phase, including the specific preparatory work that investors should undertake and the departments they should engage with. It is essential to adhere to and implement relevant national policies, laws, and regulations during this phase.

Part Ⅱ　Learning Goals

1. Master common words in oral English.Comprehend conversational English.

2. Understand the common grammar used in oral English.

3. Achieve proficiency in pronouncing English words correctly in oral communication. Develop students' oral skills.

4. Acquire relevant knowledge of the project decision phase.

5. Gain mastery of the main tasks involved in the project decision–making.

Part Ⅲ　Ideological and Political Points

1. Adhere strictly to relevant national policies, laws, and regulations.Conduct scientific analysis and demonstration.

2. Stay informed about the latest industry developments, as well as new techniques, processes, and materials.

3. Grasp the worldview and methodology of socialist ideology with Chinese characteristics in the new era.

4. Foster a continuous development of people's democracy throughout the entire process and enrich the spiritual well–being of the population.

5. Students will understand the importance of environmental protection through feasibility studies.

6. Students will recognize the need to expedite the establishment of a new development pattern and prioritize high–quality development.

7. Students will be aware of the need to accelerate the construction of trade power and promote high–quality development along "the Belt and Road" initiative.

Task 2 Oral Practice

The Main Work of Project Decision

Diana—Investor
Alex—Vice General Manager
Shrek—Project Manager
Location: Meeting room

对话音频

Diana: We plan to build a mechanical equipment factory in this city; site selection is crucial. Shrek, as the project manager, you will be responsible for overseeing the entire construction process. When preparing for the project, we should also consider the requirements of the 20th National Congress Agenda, which emphasize the promotion of a beautiful China, carbon reduction, pollution reduction, green expansion, and ecological priority, as well as green and low-carbon development.

Shrek: I will ensure more people are involved in the site selection process and related tasks.

Diana: Alex, you will take the lead in preparing the project proposal.

Alex: Yes boss, I will conduct market research immediately.

Diana: If you need any assistance, feel free to ask Shrek for help.

Alex: That would be greatly appreciated.

Shrek: Yes, boss!

Diana: It is essential that all of you have a solid understanding of the relevant national policies, laws, and regulations.

Alex: We will.

Diana: Additionally, it is crucial to keep up with the latest developments, techniques, processes, and materials in the industry.

Alex: Got it.

Shrek: I will provide Alex with full cooperation.

Secretary: This is Diana's investment conception.

Diana: Please finish it as soon as possible!

Alex: We will try our best!

Four weeks later

Shrek: Boss, the site is in Zone A.

Diana: I prefer Zone A. Prepare relevant documents for the project site-selection as soon as possible.

Alex: Diana, our project proposal has been approved by the National Development and Reform Commission.

Diana: The next step is to conduct a feasibility study. Shrek, as the technical expert, please focus more on economic and technological aspects.

Shrek: Clear!

Diana: I expect you to provide scientific analysis and a thorough demonstration. Ensure accuracy in the results.

Alex: We will!

Diana: Furthermore, arrange relevant personnel to conduct environmental assessment and safety evaluation.

Alex: I will do it at once to save time.

Diana: How long will it take?

Alex: About half a month for the report, 7–10 days for assessment, and 7 days for approval.

Diana: Please shorten the writing time on precise premise.

Alex: Got it.

Five weeks later

Alex: The project has been established, boss.

Diana: In that case, apply to the Natural Resources Bureau for the land.

Alex: It needs about 30 days.

Diana: Strive to build a green building with a rating of at least two stars, truly achieving energy conservation and emission reduction in line with national objectives.

Alex: Understood.

Diana: Once the land application is submitted, we need to find a suitable geological survey and design company promptly. This will allow us to start the preliminary design once we have obtained the land.

Shrek: We will prepare the tasks separately.

Diana: Ensure the sources of funds for construction!

Alex: Clear!

Diana: Report the progress to me every Monday.

Alex: OK!

Task 3 Analysis of Civil Construction Knowledge

Part I Related Knowledge Links

Relevant code	Content of the project decision phase	Content of the feasibility studying report	Project proposal

Part II Task Group

<table>
<tr><td>Class</td><td></td><td>Set No.</td><td></td><td>Tutor</td><td></td></tr>
<tr><td>Group Leader</td><td></td><td>Student ID</td><td></td><td></td><td></td></tr>
<tr><td rowspan="3">Team Members</td><td colspan="2">Name</td><td>Student ID</td><td>Name</td><td>Student ID</td></tr>
<tr><td colspan="2"></td><td></td><td></td><td></td></tr>
<tr><td colspan="2"></td><td></td><td></td><td></td></tr>
<tr><td>Division of Tasks</td><td colspan="5"></td></tr>
</table>

Part III Task Analysis

1. How to memorize new words quickly?
2. How to understand the meaning of the project decision?
3. Why do feasibility studying on the project?
4. What preparations are required for the project decision phase?
5. How much time is needed to evaluate the project feasibility study report?
6. Which department has the responsibility for approving the project proposal?
7. Who is in charge of obtaining land approval?

Part IV Access to Information

Leading Questions

1. What is the process for an investor to initiate a new project?

2. Which government entity is involved in the decision-making phase of the project?

3. What are the reasons for advocating energy saving and emission reduction?

4. What is the purpose of conducting feasibility analysis?

Part V Task Implementation

Leading Questions

1. What is the category of the project?

2. What is the building category of the plant?

3. Why is geological prospecting necessary for the project?

4. What preparations should be made before the project is approved?

5. Why is funding necessary for projects?

6. Why is it important to stay updated on the latest developments in the construction industry?

Part VI Task Optimization

Leading Questions

1. During the role play in groups, provide the best answer to each other.

2. Teachers and students engage in further discussions to propose revision suggestion.

3. Reflect on your feelings about the industry during the role play.

Module Two

Project Design Preparation Phase

Task 1 Task Description

Task 2 Oral Practice

Task 3 Analysis of Civil Construction Knowledge

Task 1 Task Description

Part I Task Description

Familiarize yourself with the process of preparing a design task book; understand the content that should be included in the design task book; comprehend the requirements for land use index and area index; determine building parameters; understand the requirements for Green Buildings; and coordinate the design conditions with relevant departments.

Part II Learning Goals

1. Master common words in oral English.

2. Understand the common grammar used in oral English.

3. Gain proficiency in the relevant knowledge of the project design preparation phase.Understand the preparation phase of the project.

4. Foster students' ability for self-study.Master professional knowledge through oral practice.

5. Cultivate students' teamwork spirit.Develop strong oral English skills for effective workplace communication.

Part III Ideological and Political Points

1. Foster students' craftsmanship spirit.Uphold the leadership of the CPC and adhere to socialism with Chinese characteristics.

2. Raise students' awareness about environmental and water resource protection.

3. Encourage students to fulfill their roles and responsibilities in their work.

4. Aim for a people-oriented approach. Focus on the well-being of the people's livelihood.

5. Enhance the overall level of civilization in society and implement projects for building civic morality.

6. Reusing wastewater helps to conserve water resources.

7. Promote cultural confidence and self-improvement, creating new achievements in socialist culture.

Task 2 Oral Practice

Preparation of Design Specification

Diana—Investor
Alex—Vice General Manager
Shrek—Project Manager
Vivian—Plant Manager
Titi—Civil Engineer
Bard—Electrical Engineer
Mike—Equipment Engineer
Location: Meeting room

对话音频

Diana: Now, Let us discuss the design specification. The approved land area is 40 hectares, which we will divide into three phases.

Alex: How large will the first phase of the project be?

Diana: I expect the construction area is about 80,000m^2.

Alex: Then let us start from the land use index.

Diana: Four independent workshops, one office building, one canteen, and one staff dormitory.

Alex: Vivian, please give your requirements.

Vivian: I will give you a detailed description after the meeting.

Diana: Cost saving, while meeting the design requirements.

Shrek: I will list the requirements of the general layout.

Alex: The design profundity will be marked clearly.

Titi: I will take charge of the design requirements of civil engineering.

Bard: I will contact the State Grid to determine power load conditions and design scope.

Mike: I will handle the requirements from the Water Company and the Heating Company. I will do my best to gather the basic design conditions.

Alex: Shrek, do not forget the geological investigation report.

Shrek: I will give it to the Design Company with specification.

Alex: Vivian, think about ground parking space index. Shrek will cooperate with you referring to general drawing.

Vivian: I suggest parking near the main entrance intensively.

Shrek: Got it.

Diana: Putting our staff first, we aim for a people-oriented approach!

Vivian: OK!

Diana: Alex, you are responsible for Green Buildings and energy saving requirements.

Alex: Following the national standards, hope to exceed the national standards.

Diana: Shrek, do a good job in cost analysis and control.

Alex: Mike, get to know the requirements of our project for sponge city. Strive towards the goal of "deepening the prevention and control of environmental pollution, and continuously fighting the defense of blue sky, clear water, and pure land" in the spirit of the 20th National Congress.

Mike: I will.

Diana: Regarding the design period, I propose 25 days for the concept design, 25 days for the preliminary design, and 50 days for the construction drawings. Is this feasible?

Alex: I will finalize it with the Design Company again.

Diana: That is all for now, details of the design assignment will be prepared separately.

Alex: OK, we will complete the first draft before next Monday!

Task 3 Analysis of Civil Construction Knowledge

Part I Related Knowledge Links

Green Building	Boundary line of land	Sponge city	General layout	Land area	Design profundity

Part II Task Group

<table>
<tr><td>Class</td><td></td><td>Set No.</td><td></td><td>Tutor</td><td></td></tr>
<tr><td>Group Leader</td><td></td><td>Student ID</td><td></td><td></td><td></td></tr>
<tr><td>Team Members</td><td colspan="5"><table>
<tr><td>Name</td><td>Student ID</td><td>Name</td><td>Student ID</td></tr>
<tr><td></td><td></td><td></td><td></td></tr>
<tr><td></td><td></td><td></td><td></td></tr>
</table></td></tr>
<tr><td>Division of Tasks</td><td colspan="5"></td></tr>
</table>

Part III Task Analysis

1. What preparations are necessary for the project design preparation phase?
2. What is the intended scale of the construction project?
3. What is the concept of Green Buildings?
4. What is the proposed plan for the design period?
5. What preparations are necessary before starting the design phase?
6. Why is it important for the project to meet the national standards?
7. Why do we need to obtain geological survey reports?

Part IV Access to Information

Leading Questions

1. Why is it important for the project to meet national specifications?

2. What are the differences between the roles of the investor and the plant manager?

3. What are the reasons for promoting "sponge city" ?

4. What is the design task book?

Part V Task Implementation

Leading Questions

1. How many buildings are there in the plant?

2. What is meant by the term "sponge city" ?

3. Why is it necessary to determine the design and construction period in advance?

4. Please assign a design task book for this project to other teams.

5. How do you control costs in this project?

6. What are the requirements for municipal conditions in the installation project?

Part VI Task Optimization

Leading Questions

1. How to obtain the relevant conditions of power supply?

2. Summarize the engineering content of this unit.

Module Three

Project Preliminary Designing Phase (Ⅰ)

Task 1　Task Description
Task 2　Oral Practice
Task 3　Analysis of Civil Construction Knowledge

Task 1 Task Description

Part I Task Description

Understand the format of scheme design; identify the professions involved in the project completion; comprehend the purpose of comparing plans among different disciplines; and recognize the significance of cost analysis.

Part II Learning Goals

1. Master common words in oral English.
2. Understand the common grammar used in oral English.
3. Develop professional skills through oral practice.
4. Acquire professional knowledge and enhance cognitive clarity.
5. Master professional knowledge through oral practice.
6. Understand the preparation phase of the project.
7. Expand students' knowledge and understandings.
8. Cultivate students' innovative spirit.

Part III Ideological and Political Points

1. Value job opportunities and dedicate yourself wholeheartedly to your work.
2. Promote the dissemination and influence of Chinese civilization and uphold the position of Chinese culture.
3. Build a strong professional foundation for your work.
4. Share positive stories about China and amplify the voice of China.
5. Achieve cost savings while ensuring construction quality.
6. Uphold the application of dialectical materialism and historical materialism.
7. Consider issues from the perspective of investors.
8. Enhance people's well–being and improve their quality of life.

Task 2 Oral Practice

The Concept Design

Jodie—Design Manager
Holger—Civil Engineer
Divid—Electrical Engineer
Sam—Equipment Engineer

对话音频

Jodie: Did all of you read the design specification?

Holger: Yes, we did.

Jodie: Any doubts?

Divid: No.

Jodie: Holger, please arrange for the scheme designer to start.

Holger: No problem.

Jodie: For the architecture, structure, electric, HVAC, water, and other professions, you should prepare the concept design before the preliminary design.

Sam: Should we just refer to the design specification?

Jodie: More than that, we should also consider the general layout and the area index provided by the investor.

Holger: I have no idea.

Jodie: Based on the past experience, we can almost be certain about the design contents for such a large plant.

Divid: Only concept design in text?

Jodie: Yes, I will list design contents and provide at least two options for each section and system of each specialty.

Sam: Will we compare the different options?

Jodie: Yes, we will provide a final choice suggestion.

Holger: I will supervise the scheme designer to provide us with a rough plan as soon as possible.

Jodie: Don't miss anything.

Divid: It sounds a bit difficult.

Jodie: A good designer should always keep the design content in mind for different types of buildings. We should adhere to the concept of the 20th National Congress of the Communist Party of China: "Prosperity and development of cultural undertakings and industries require a people-centered creative orientation and the

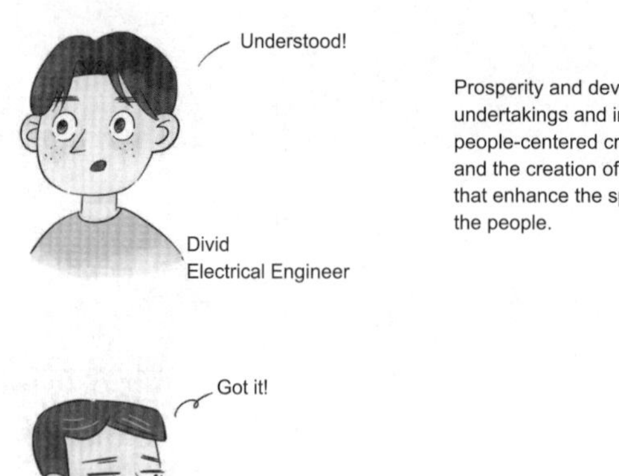

creation of excellent works that enhance the spiritual strength of the people."

Sam: Got it!

Jodie: Also, pay attention to cost analysis.

Holger: What? Are we going to do the cost estimation?

Jodie: You can think so.

Divid: Should not it be done by the investor?

Jodie: A professional designer should do more than just design.

Sam: It is challenging!

Holger: Does every solution come with a cost?

Jodie: Right. It will be convenient for investors to make a choice.

Sam: Does it mean we do the preliminary design after the investor confirms the system plan?

Jodie: Yes. The concept design only takes 25 days. Do not spend too much time on this part of the work.

Holger: Clear!

Jodie: I want to see the summary of all professionals on Friday. I hope you will fully embrace the spirit of craftsmanship and constantly strive for perfection.

Divid: OK.

Task 3 Analysis of Civil Construction Knowledge

Part Ⅰ Related Knowledge Links

The concept design	Cost analysis	Differences between estimation and budget	The spirit of craftsmanship	Plan of concept

Part Ⅱ Task Group

<table>
<tr><td>Class</td><td></td><td>Set No.</td><td></td><td>Tutor</td><td></td></tr>
<tr><td>Group Leader</td><td></td><td>Student ID</td><td></td><td></td><td></td></tr>
<tr><td rowspan="3">Team Members</td><td>Name</td><td>Student ID</td><td>Name</td><td colspan="2">Student ID</td></tr>
<tr><td></td><td></td><td></td><td colspan="2"></td></tr>
<tr><td></td><td></td><td></td><td colspan="2"></td></tr>
<tr><td>Division of Tasks</td><td colspan="5"></td></tr>
</table>

Part Ⅲ Task Analysis

1. What is the concept design?
2. Why is it important to compare different options during the design process?
3. What is the purpose of conducting cost analysis during the concept design phase?
4. What does a main scheme comparison entail?
5. What are the investors' primary concerns?
6. Why is it important to design the general layout?

Part Ⅳ Access to Information

Leading Questions

1. What are the essential professional qualities that a qualified designer should possess?

2. Why is it important for designers to have an understanding of cost analysis?

3. How to consider the relationship between cost and quality?

4. In your opinion, which aspect do investors prioritize more, cost or quality?

Part V Task Implementation

Leading Questions

1. What components should be included in the cost estimation for this project?

2. Which professionals are involved in the design phase of the project?

3. As an investor, what factors would influence your selection of a scheme?

Part VI Task to Optimization

Leading Questions

1. What are the key components of a professional design project?

2. How can we facilitate group discussions on cost-saving measures?

3. How can we effectively address the needs of investors through group discussions?

Module Four

Project Preliminary Designing Phase（Ⅱ）

Task 1　Task Description
Task 2　Oral Practice
Task 3　Analysis of Civil Construction Knowledge

Task 1 Task Description

Part I Task Description

Understand the design depth of preliminary design; familiarize yourself with the importance of preliminary design; be acquainted with the required drawing content for each profession; and ensure that the preliminary design complies with relevant national laws and regulations.

Part II Learning Goals

1. Master common words in oral English.
2. Understand the common grammar used in oral English.
3. Master knowledge in architecture.
4. Enhance skills and abilities.
5. Master the basis of the design.
6. Master the contents of the preliminary design.

Part III Ideological and Political Points

1. Establish a strong theoretical foundation and possess insightful thinking.
2. Uphold the leadership of the Party and prioritize people as the masters.
3. Develop a sense of initiative and collaboration.
4. Encourage diverse ideas and open discussions.
5. Strictly comply with national laws and regulations.
6. Fully implement the responsibility system for ideological work.
7. Prioritize quality in the entire project.Emphasize the importance of teamwork.
8. Consolidate, strengthen, and advance the mainstream ideology and public opinion of the New Era.

Task 2 Oral Practice

Preliminary Design

Jodie—Design Manager
Holger—Civil Engineer
Divid—Electrical Engineer
Sam—Equipment Engineer

对话音频

Jodie: After several modifications, the scheme has been approved by the Planning Bureau. This is the feedback on the concept design from investors. Now, Let us discuss the preliminary design.

Holger: Is the preliminary design reviewed by experts or by the investors internally?

Jodie: Of course, it is reviewed by the experts.

Divid: It means the design must meet the local requirements.

Sam: Some requirements in the design specification may change.

Jodie: If there are any problems, contact the investor immediately.

Sam: OK!

Jodie: Now, let us report the content of preliminary design separately.

Holger: For the architecture professionals, it includes the design specification, general plan, plans for each level, elevations, and sections. Structure professionals should include the design specification, pile positions and foundation plans, beam and plate drawings for each layer, constructional requirements of new structures, and diagram nodes. That is all!

Divid: For the electric professionals, it includes the design specification, general plan of power supply, transformer substation and distribution station, power plan, system diagram, building lightning protection, weak current system diagram, main equipment, and material list. That is all!

Sam: For the HVAC professionals, it includes the design specification, air conditioning and ventilation plan, primary machine room, heat exchange station, heating, main equipment, and material list. Water supply and drainage professionals should include the design specification, pipe general layout, pipe plan for each level, water supply system, drainage system, fire pipe system, main equipment, and material list. That is all!

Jodie: Okay, make sure not to miss any item. Be detailed enough and cooperate well with each other. Do not make any fundamental mistakes and do not violate any mandatory provisions of the code. We must adhere to the principles of law–based governance and develop law–based governance in China. We must also adhere to the path of socialist rule of law with Chinese characteristics.

Holger: Got it!

Jodie: Concentrate on the design. It is only 25 days. Maybe we have to work overtime. Thanks for your hard work.

Divid: We will do what we can!

Task 3 Analysis of Civil Construction Knowledge

Part Ⅰ Related Knowledge Links

State Grid	Air conditioning	Fire system	Compulsory provision

Part Ⅱ Task Group

<table>
<tr><td>Class</td><td></td><td>Set No.</td><td></td><td>Tutor</td><td></td></tr>
<tr><td>Group Leader</td><td></td><td>Student ID</td><td></td><td></td><td></td></tr>
<tr><td rowspan="3">Team Members</td><td>Name</td><td>Student ID</td><td>Name</td><td colspan="2">Student ID</td></tr>
<tr><td></td><td></td><td></td><td colspan="2"></td></tr>
<tr><td></td><td></td><td></td><td colspan="2"></td></tr>
<tr><td colspan="6">Division of Tasks</td></tr>
</table>

Part Ⅲ Task Analysis

1. What is the preliminary design?
2. Why is the preliminary design significant?
3. How can the initial design benefit the investor?
4. Do different professions have the same contents in the preliminary design?
5. What are the compulsory provisions?
6. Which holds greater importance, national standards or local requirements?

Part Ⅳ Access to Information

Leading Questions

1. What aspects will the water professionals prepare for the preliminary design?

2. What aspects will the HVAC professionals focus on during the preliminary design?

3. What is the practice of lightning protection and grounding?

4. What effect do transformer substations and distribution stations serve?

Part V Task Implementation

Leading Questions

1. Who is responsible for deciding the location of the transformer substation and distribution station?

2. Is there a difference between the weak current system and the intelligent system?

3. Do you think the plan or the elevation should be drawn first? Why?

4. What information should be included in the material sheet?

Part VI Task Optimization

Leading Questions

What is the interrelationship between different specialties?

Module Five

Construction Drawing Design Phase（Ⅰ）

Task 1 Task Description
Task 2 Oral Practice
Task 3 Analysis of Civil Construction Knowledge

Task 1 Task Description

Part I Task Description

Familiarize yourself with the design content included in the complete project construction drawing; understand how to collaborate with other disciplines in civil engineering to complete the construction drawing; determine whether each discipline needs to provide design condition drawings to other disciplines; and establish a reasonable timeline for the project milestones.

Part II Learning Goals

1. Master the common words in oral English.

2. Understand the common grammar used in oral English.

3. Master the concept of the construction drawing design.

4. Develop effective collaboration skills among professionals.

5. Familiarize yourself with the role of relevant municipal authorities in the project.

6. Recognize the importance of pre–burying and reservations.

Part III Ideological and Political Points

1. Effective work division and cooperation can lead to exemplary projects. Possess keen insight.

2. Focus on cultivating new generations responsible for national rejuvenation.

3. Foster empathy and cultivate a spirit of selfless dedication.

4. Prioritize the direct and realistic interests of the people.

5. Reflect the will of the people, safeguard their rights and interests, and stimulate their creative vitality.

6. Enhance communication and problem–solving skills. Ensure the reliability of data and information.

7. Accelerate the development of a modern economic system and focus on improving total factor productivity.

Task 2 Oral Practice

Design Dote Discussion

Jodie—Design Manager
Holger—Civil Engineer
Divid—Electrical Engineer
Sam—Equipment Engineer

对话音频

Jodie: According to the preliminary design review, we need to make some changes to the plan and start the construction drawing design. The content of the construction drawings aligns with the call of the 20th National Congress of the Communist Party of China: "Adhere to the principle of standing first and then breaking and implement the carbon peak action with a plan and steps."

Holger: Now, Let us set up design note time.

Jodie: When will you show them the architectural condition drawing?

Holger: At least one week later.

Jodie: Including foundation drawing?

Holger: No, that will need more time.

Sam: You can give us the floor plan first.

Divid: It would be better to have the beam drawing earlier than the foundation drawing.

Holger: We will try our best.

Jodie: Give them the floor plan one by one to avoid making them wait too long.

Holger: OK, we will provide workshop 1 drawing first.

Sam: I still need the municipal interface information.

Jodie: I will contact the relevant departments later.

Divid: It would be better if the Construction Company could personally contact the relevant municipal departments.

Jodie: I will inform them.

Holger: Complete architectural floor plans will be given to all of you in two weeks.

Sam: The reserved hole and load information on the floor will be provided along with that.

Holger: OK.

Divid: We can provide the preliminary conditions of the Distribution Room, but the details need to be provided by the State Grid.

Jodie: Reserve the space first, Divid.

Divid: Yes.

Holger: Sam, you should start thinking about the space for the pump room and the heat exchange station now.

Sam: Got it. But I am also waiting for water pressure information from the Water Company.

Divid: When will you give me the equipment load, Sam?

Sam: One week later.

Jodie: Provide the design requirements to the relevant professional first to avoid affecting the working progress of others. The Equipment Professional Department should try to use energy-saving products.

All: Yes.

Task 3 Analysis of Civil Construction Knowledge

Part I Related Knowledge Links

Construction drawing design	Drawing review	Architectural drawing	Heating Company	Water Company	Transformer distribution station

Part II Task Group

Class		Set No.		Tutor	
Group Leader		Student ID			
Team Members	Name	Student ID	Name	Student ID	
Division of Tasks					

Part III Task Analysis

1. What is the construction drawing design?
2. Is each professional responsible for designing independently?
3. Why is it important to review construction drawings?
4. Why do we need to do pre-burying and reservation in civil design?
5. Why do we need get water pressure information from the Water Company?
6. Why should the structural professional determine the load information?

Part IV Access to Information

Leading Questions

1. Why is it necessary for architectural professionals to provide design conditional drawings for other professionals?

2. Why should the structural foundation drawing be provided to other professionals first?

3. What design information will we get from the Heating Company?

4. What design information will we get from the Water Company?

Part V Task Implementation

Leading Questions

1. What are the methods used for reviewing preliminary designs?

2. What are the differences between the design depth of the construction drawing and the preliminary design?

3. As a designer, what skills do you believe are essential for successful design?

4. If you are in charge of design, how do you plan your schedule?

Part VI Task to Optimization

Leading Questions

1. What is the function of the pump house?

2. Summarize the key issues to consider in the design process for each profession.

Module Six

Construction Drawing Design Phase（Ⅱ）

Task 1 Task Description

Part I Task Description

Understand which individual buildings are included in the project's construction drawing; familiarize yourself with the design parameter requirements for each building; ensure that the design depth of the construction drawing meets the requirements of the drawing review; determine the locations for equipment installation facilities; and familiarize yourself with the standards for Green Buildings.

Part II Learning Goals

1. Master common words in oral English.
2. Understand the common grammar used in oral English.
3. Understand the significance of Green Buildings.
4. Familiarize yourself with the parameters of prefabricated buildings.
5. Familiarize yourself with the standards of the construction drawing design.
6. Understand the different categories of buildings.

Part III Ideological and Political Points

1. Advocate the concept of green and low-carbon. Avoid extravagance and waste.

2. Prioritize talent as the primary resource and innovation as the primary driving force.

3. Emphasize the promotion of governance and management according to law.

4. Utilize advanced technology and processes. Adopt a rigorous scientific attitude.

5. Emphasize the importance of using qualified building materials.

6. Recognize the impact of errors, leaks, and collisions on construction quality.

7. Highlight the significance of ingenious design in creating high-quality projects.

8. Stress the application of dialectical materialism and historical materialism.

9. Unity is strength.

Task 2 Oral Practice

Discussion of Architectural Design Details

Holger—Civil Engineer
Alvin—Architectural Designer
Olay—Architectural Designer
Peter—Architectural Designer

对话音频

Holger: Now, let us confirm the design.

Alvin: Please give me a building design that requires less work. I have limited abilities.

Holger (with a smile): Don't worry, we trust you.

Olay: The requirements for reviewing drawings are becoming more stringent.

Peter: So, we need to comprehend the design specification thoroughly.

Holger: Peter will be responsible for the workshop's design, Olay for the office building and canteen, and Alvin will assist Peter in designing the dormitory and workshop. The requirement is to embody prefabricated buildings and achieve the overall goal of "adhering to the focus of economic development on the real economy, promoting new industrialization, and accelerating the construction of a strong manufacturing and quality country" in the spirit of the 20th National Congress.

Olay:Which individual building will be the prefabricated building?

Holger: Dormitory.

Alvin: It sounds a bit complicated.

Holger: Come on, you can handle it.

Olay: It is a piece of cake for you.

Holger: The exterior wall coating will be made of real lacquer.

Peter: All of them?

Holger: Yes, except for the workshops, which will have steel structures. Internal drainage will be used on the roof.

Peter: I will inform the water designer to supply the size of the roof gutter.

Holger: After receiving it, we will promptly provide feedback to the steel Structure Company for review.

Peter: I will.

Holger: You should have more professional communication with the structural

engineer regarding the assembly design. The parameters are very strict.

Alvin: I understand.

Holger: Some parameters should be determined by the assembly manufacturer.

Alvin: OK.

Holger: You need to think about the central air conditioning and sprinkler system when determining the height of the office building and canteen.

Olay: What is the owner's requirement on effective height?

Holger: It is no less than 2.8m for the using room, and 2.4m for toilets.

Olay: Should we consider the ceiling?

Holger: Yes. By the way, the workshop will have fans and solar panels on the roof, and the position of the hot water tank will be determined by the water designer. The shower room will be placed in workshop 1, and shower facilities in the dormitory should also be taken into consideration.

Peter: OK.

Holger: Each building should meet the two–star standard and comply with Green Building requirements. So, that is it. I will discuss the details of each building with each designer individually.

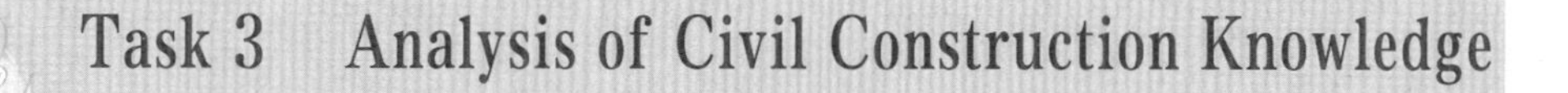

Task 3 Analysis of Civil Construction Knowledge

Part Ⅰ Related Knowledge Links

Sprinkler system	Steel structural building	Central air conditioning system	Prefabricated building	Real lacquer

Part Ⅱ Task Group

<table>
<tr><td>Class</td><td></td><td>Set No.</td><td></td><td>Tutor</td><td></td></tr>
<tr><td>Group Leader</td><td></td><td>Student ID</td><td></td><td></td><td></td></tr>
<tr><td>Team Members</td><td colspan="5">
<table>
<tr><td>Name</td><td>Student ID</td><td>Name</td><td>Student ID</td></tr>
<tr><td></td><td></td><td></td><td></td></tr>
<tr><td></td><td></td><td></td><td></td></tr>
</table>
</td></tr>
<tr><td>Division of Tasks</td><td colspan="5"></td></tr>
</table>

Part Ⅲ Task Analysis

1. How to determine the star standard of Green Buildings?
2. Which buildings require an automatic spraying system?
3. Why is there a limitation on the effective height of buildings?
4. Why do we need to consider the central air conditioning and sprinkler system in the office building and canteen?
5. What factors determine the parameters of the air conditioning system?
6. Can you explain the principle of solar water heating?

Part Ⅳ Access to Information

Leading Questions

1. What role do design codes play in the engineering design?

____

2. What are solar panels used for?

3. How does the load of the roof water tank affect the floor slab?

4. What are the differences between rainwater internal drainage and external drainage?

Part V Task Implementation

Leading Questions

1. Can you describe the differences in equipment installation between the canteen and the dormitory?

2. Can you explain the working principles of the air conditioning system?

3. What happens if the design load is less than the actual load?

4. Can you name other heat sources besides solar energy? What are they?

Part VI Task to Optimize

Leading Questions

1. What are the standards of two-star Green Building?

2. Describe your first impression on a prefabricated building.

Module Seven

Construction Drawing Design Phase (Ⅲ)

Task 1 Task Description
Task 2 Oral Practice
Task 3 Analysis of Civil Construction Knowledge

Task 1　Task Description

Part I　Task Description

Understand different types of building foundations; understand the seismic fortification intensity requirements for buildings; familiarize yourself with the structural forms of each building; and ensure that the design strength and load meet the corresponding specification requirements.

Part II　Learning Goals

1. Master common words in oral English.
2. Understand the common grammar of oral English.
3. Master the classification of seismic grade.
4. Master the type of building foundation.
5. Familiarize yourself with different structure types.
6. Gain a comprehensive understanding of fabricated building technology.

Part III　Ideological and Political Points

1. Prioritize people's safety and establish a correct value system.
2. Recognize the impact of energy conservation and emission reduction on the environment.
3. Uphold the principles of high-quality development and the strategy of national revitalization through science and education.
4. Promote mutual respect.
5. Foster global friendship.
6. Recognize the importance of faith in one's life.

Task 2 Oral Practice

Discussion of Structural Design Details

Holger—Civil Engineer
Bruce—Structural Designer
Nate—Structural Designer
Leona—Structural Designer

对话音频

Holger: Let us discuss some details now. Nate will be responsible for the workshop design, Bruce for the office building and canteen, and Leona for the dormitory.

Bruce: I want to confirm if the investor has chosen independent or pile foundation.

Holger: Independent foundation with a waterproof board.

Nate: Will there be a crane beam in the workshop?

Holger: Yes, in workshop 2 and workshop 4. The load shall be provided by the investor.

Leona: Integral assembly type for dormitory?

Holger: Yes, it is. The floor slab will consist of truss reinforced concrete composite slab with a cast-in-place layer. The bearing walls will be in the form of in-situ shear walls. However, the vertical components have not adopted assembly technology.

Nate: Will steel structure be used in all workshops?

Holger: Yes. The buildings have a seismic fortification intensity of 7 degrees, adhering to the requirements of the 20th National Congress of the Communist Party of China, which emphasizes comprehensive national security. Rigorous calculations are needed to control the amount of steel.

Bruce: Will the office building and canteen use frame and shear structure?

Holger: The form of the staircase will be recast reinforced concrete slab with cast-in-place platform for all buildings. All the steel bars used in this project will be high strength steel bars. The design load must meet the specifications, and attention should be paid to some parameters required for Green Buildings.

Nate: Got it.

Adhering to the requirements of the 20th National Congress of the Communist Party of China.
Holger
Civil Engineer

Task 3 Analysis of Civil Construction Knowledge

Part I Related Knowledge Links

Crane beam	Cast-in-place	Seismic intensity	Steel type	Frame and shear structure	Truss

Part II Task Group

Class		Set No.		Tutor	
Group Leader		Student ID			

Team Members	Name	Student ID	Name	Student ID
Division of Tasks				

Part III Task Analysis

1. How do you determine the type of building foundation?

2. Can you explain the differences between assembly and cast-in-place construction?

3. What is the function of the crane beam?

4. How can seismic zones be distinguished?

5. How can seismic zones be marked?

6. Are you familiar with strip foundations?

Part IV Access to Information

Leading Questions

1. What are the differences between independent foundations and pile foundations?

2. How familiar are you with prefabricated buildings?

3. In what types of buildings are truss systems primarily used?

4. Can masonry be used for high-rise buildings?

Part V Task Implementation

Leading Questions

1. What is high-strength reinforcement?

2. Please describe the structural form of buildings.

3. What is ordinary reinforcement?

4. Do you know the laminate plate technology?

Part VI Task to Optimization

Leading Questions

1. What is the basis for determining seismic intensity?

2. What does the main structure of a building include?

Module Eight

Construction Drawing Design Phase（Ⅳ）

Task 1　Task Description
Task 2　Oral Practice
Task 3　Analysis of Civil Construction Knowledge

Task 1 Task Description

Part Ⅰ Task Description

Understand the fire water design requirements for different types of building; familiarize yourself with the fire hazard level of the warehouse in the project; understand the design requirements for the strength and functional area of sprinkler systems. If the pressure of the municipal water supply does not meet the fire pressure requirement, it is necessary to design a fire water reservoir. Determine the type of sprinkler system used in the warehouse.

Part Ⅱ Learning Goals

1. Master common words in oral English.
2. Understand the common grammar used in oral English.
3. Acquire knowledge of fire fighting basics.
4. Understand the classification of fire protection systems.
5. Master the parameters required.
6. Understand the importance of fire prevention.

Part Ⅲ Ideological and Political Points

1. Develop awareness of safety and fire prevention.
2. Prioritize the principle of valuing life first.
3. Possess basic knowledge of fire escape procedures.
4. Propose and implement new development philosophies, promoting the construction of a new development pattern.
5. Develop a correct outlook on life.
6. Have clear targets and be devoted to work.
7. All for one, one for all.
8. Adhere to the principles of a high degree of autonomy.

Task 2 Oral Practice

Fire Water System Design Details—Parameter Determination

Sam—Equipment Engineer
Andy—Water Designer
Eric—Water Designer

对话音频

Sam: As a class E workshop, the fire water consumption indoors and outdoors needs to strictly follow the requirements of the Technical Code for Fire Protection Water Supply and Hydrant Systems.

Andy: The difficulty lies in the sprinkler systems in the warehouse.

Sam: That is exactly what we need to focus on.

Eric: As it is a class E facility, why is a sprinkler system needed in the warehouse?

Sam: Because the raw materials are stored in the warehouse with wooden packing boxes.

Andy: According to the code for design of sprinkler systems, the warehouse has a fire danger level of Level Ⅱ. The spraying intensity is 12L/(min · m^2), the action area is 200m^2, the continuous spraying time is 1.5 hours, and the final fire flow is 40L/s.

Sam: However, the investor demands a spraying intensity of 16L/(min · m^2) and an action area of 240m^2. So you need to recalculate the flow again.

Eric: That much flow! This means the volume of the fire reservoir has to be increased.

Sam: Yes, we should place the fire pool and pump room on the ground floor of the office building. The municipal water pressure is only 0.25MPa with a single pipeline for supply, so the volume of the fire reservoir must meet the requirements for indoor and outdoor fire water.

Eric: Got it.

Andy: If the investor doesn't have any special requirements, we will adopt the Early Suppression Rapid Response (ESFR) type sprinklers, which are specially designed for warehouses according to the code. This ensures the fastest protection of life and property in the event of a fire, in line with the spirit of the 20th National

Congress of the Communist Party of China, which emphasizes ensuring people's safety and taking political and economic safety as priorities.

Sam: We just need to follow the code.

Eric: Is the fire water reservoir on the roof of the office building?

Sam: Which one is higher, dormitory or office building?

Eric: Same.

Sam: In that case, we will place it on the roof of the office building because it is closer to the pump room.

Eric: OK!

Sam: Anything else? By the way, we will use a siphon rainwater system for drainage inside the roof.

Eric: No more questions.

Sam: That is all for today. Please continue with the drawings.

Task 3 Analysis of Civil Construction Knowledge

Part Ⅰ Related Knowledge Links

Hydrant system	Fire pump room	Siphon rainwater	Roof fire water reservoir	Sprinkler

Part Ⅱ Task Group

Class		Set No.		Tutor	
Group Leader		Student ID			

Team Members	Name	Student ID	Name	Student ID

Division of Tasks	

Part Ⅲ Task Analysis

1. How to extinguish a fire with a sprinkler system?
2. Why is a warehouse equipped with a sprinkler system?
3. Can you explain the principle of sprinkler systems?

Part Ⅵ Access to Information

Leading Questions

1. How many automatic factors do sprinkler systems have?

2. How do we calculate the volume of a fire reservoir?

3. What are the differences between a siphon rainwater system and a gravity flow rainwater system?

4. How is the size of a roof water reservoir determined?

Part V Task Implementation

Leading Questions

1. What is a siphon rainwater system?

2. Is it safe to use elevators in the event of a fire?

3. What is the function of municipal water pressure?

4. What materials are used for a roof water reservoir?

Part VI Task to Optimize

Leading Questions

1. Can non-firefighters operate the fire hydrant?

2. How does a fire truck operate?

Module Nine

Construction Drawing Design Phase（V）

Task 1　Task Description

Task 2　Oral Practice

Task 3　Analysis of Civil Construction Knowledge

Task 1 Task Description

Part I Task Description

Understand which buildings utilize centralized air conditioning systems and which buildings utilize heating systems; be familiar with the types of municipal heat sources; understand the configuration of air conditioning systems; and have knowledge of the concept of constant temperature and humidity.

Part II Learning Goals

1. Master common words in oral English.
2. Understand the common grammar used in oral English.
3. Master the importance of choosing energy-efficient sources.
4. Identify different types of heat energy sources.
5. Understand the importance of maintaining humidity in indoor environments.
6. Understand the benefits of using air source heat pumps.

Part III Ideological and Political Points

1. Creating Green Buildings requires adherence to the concept of energy conservation.
2. Uphold a deep sense of patriotism.
3. Continuous practice and theoretical innovation have no limits.
4. The path to modernization in China aims to achieve common prosperity for all people.
5. Prioritize the well-being of people and their lives.
6. Implement rules and guidelines.
7. Grassroots democracy is a critical aspect of the overall democratic process.
8. Promote the spirit of socialist rule of law and inherit the excellent traditional legal culture of China.

Task 2 Oral Practice

Air Conditioning System Design

Sam—Equipment Engineer
Jacy—Heating and air conditioning Designer
Pope—Heating and air conditioning Designer

对话音频

Sam: The office building and canteen will require a central air conditioning system. The workshops and dormitory will have a heating system.

Jacy: Where will the heat exchange station be located?

Sam: It will be next to the fire pump room on the ground floor of the office building for convenient management.

Jacy: Will the municipal heat source be hot water or steam?

Sam: It will be steam. Remember to consider the exhaust and ventilation system in the workshops, and they need to maintain constant temperature and humidity.

Pope: Okay, where will the fans be located? On the roof or on the side wall?

Sam: They will be located on the roof. The precise location and load requirements will be provided to the structural designer in a timely manner.

Pope: What are the requirements for maintaining constant temperature and humidity?

Sam: There is a document detailing the requirements. You can ask Holger for it.

Pope: Will the office building use a multi-split air conditioning system?

Sam: No, it will use a central air conditioning system.

Pope: Will it be an air source heat pump?

Sam: Yes. It will meet the energy-saving requirements. Also, don't forget to discuss the location of the air conditioning room with the architectural designer.

Pope: OK!

Sam: Pay attention to the design requirements from the investor regarding air flow, temperature and humidity.

Pope: Got it, and I will keep in touch with them if there are any questions.

Sam: Great! That is all for now. Feel free to discuss any questions in the We-Chat group. I hope everyone can innovate in your design, and the spirit of the

20th National Congress encourages us to "strengthen the dominant position of enterprises in scientific and technological innovation and exert the leading and supporting role of technology–based backbone enterprises".

Jacy: OK!

Task 3 Analysis of Civil Construction Knowledge

Part Ⅰ Related Knowledge Links

Air source heat pump	Heat exchange station	Air conditioning room	Multi-split	Constant temperature and humidity

Part Ⅱ Task Group

Class		Set No.		Tutor	
Group Leader		Student ID			
Team Members	Name	Student ID	Name	Student ID	
Division of Tasks					

Part Ⅲ Task Analysis

1. How do you select a heating system?
2. What is the function of the roof fan on a building?
3. What are the differences between an air source heat pump and solar energy?
4. How is air volume controlled?
5. Describe the working principle of an air conditioning room?
6. What is the function of maintaining constant temperature and humidity?

Part Ⅳ Access to Information

Leading Questions

1. How can energy-saving products be identified?

2. How can the vibration of a roof fan be reduced?

3. How can constant temperature and humidity be achieved?

4. Why is it necessary to calculate indoor air volume?

Part V Task Implementation

Leading Questions

1. Why should coal-fired boilers not be chosen as heat sources?

2. What is the function of a heat transfer station?

3. What are the differences between centralized air conditioning and wall air conditioning?

4. Can air conditioning units be installed anywhere?

Part VI Task to Optimize

Leading Questions

1. What is the role of air temperature in a heating system?

2. What types of buildings require centralized air conditioning?

Module Ten

Construction Drawing Design Phase (Ⅵ)

Task 1 Task Description

Part I Task Description

Determine the voltage and installation method of the municipal power supply; ensure that the location of the transformer and distribution room meets the requirements of the State Grid and local regulations; familiarize yourself with the requirements for the floor height and area of the transformer and distribution room; and be responsible for completing the detailed design of the transformer and distribution room.

Part II Learning Goals

1. Master common words in oral English.
2. Understand the common grammar used in oral English.
3. Aim knowledge about power generation.
4. Master electricity safety knowledge.
5. Master the types of electrical load.
6. Develop a sense of electricity conservation.

Part III Ideological and Political Points

1. Value and conserve national electric energy resources.
2. Understand basic electricity safety guidelines.
3. Strengthen the national spirit to achieve the great rejuvenation of the Chinese nation.Promote the core socialist values.
4. Recognize the importance of utilizing energy-saving methods in power generation.
5. Have a sense of caring for the family and everyone.
6. Young people should firmly listen to the Party's guidance and follow its lead.
7. Unity is strength. Unity is the key to success.

Task 2 Oral Practice

Who will Design Transformer and Distribution Room

Divid—Electrical Engineer
Amy—Electrical Designer
Cute—Electrical Designer

对话音频

Divid: The power consumption index requires the introduction of two 10kV power supplies from the municipal system. The 10kV power cable will be laid through a cable tray and led to the Transformer and Distribution Room. When introducing municipal power, whether it is overhead or buried, we should fully understand the spirit of the 20th National Congress of the Communist Party of China, which emphasizes the importance of respecting, conforming to, and protecting nature in building a socialist modernized country.

Amy: Will the power transformation and distribution substation be located in the underground floor of the office building?

Divid: The location should be determined with the architecture designer and sufficient space should be reserved according to the requirements of the State Grid and applicable codes.

Amy: I need to communicate with the State Grid again.

Divid: Do it as soon as possible.

Cute: Will we be responsible for designing the Transformer and Distribution Room?

Divid: We will only provide the configuration and building conditions to the architectural designer. The detailed design will be completed by the Electric Power Design Institute.

Cute: OK!

Amy: I will confirm with the State Grid about the design conditions in the afternoon.

Divid: Better to go along with the architectural designer.

Amy: OK!

Divid: Discuss with the architecture designer to determine the location of the Transformer and Distribution Room, especially regarding the floor height and area.

Amy: No problem!

Divid: On the premise of meeting the design requirements, try your best to save costs. Also, make sure it is not too far away from the municipal interface.

Amy: Got it!

Task 3 Analysis of Civil Construction Knowledge

Part Ⅰ Related Knowledge Links

Power load	Voltage level	Power generation	Principles of solar power generation

Part Ⅱ Task Group

<table>
<tr><td>Class</td><td></td><td>Set No.</td><td></td><td>Tutor</td><td></td></tr>
<tr><td>Group Leader</td><td></td><td>Student ID</td><td></td><td></td><td></td></tr>
<tr><td rowspan="3">Team Members</td><td>Name</td><td>Student ID</td><td>Name</td><td colspan="2">Student ID</td></tr>
<tr><td></td><td></td><td></td><td colspan="2"></td></tr>
<tr><td></td><td></td><td></td><td colspan="2"></td></tr>
<tr><td>Division of Tasks</td><td colspan="5"></td></tr>
</table>

Part Ⅲ Task Analysis

1. What factors need to be considered in the design of the electrical distribution room?
2. What are the requirements for connection to the municipal power supply?
3. How many cable installation methods are discussed in this dialogue?
4. Why is the detailed design completed by the Electric Power Design Institute?
5. Name the types of energy-saving power generation you are familiar with.
6. Why should be the Transformer and Distribution Room not too far away from the municipal interface?

Part IV Access to Information

Leading Questions

1. What is the function of the electrical distribution room?

2. Why does the municipal government introduce two power sources?

3. What are the benefits of using energy-saving methods for power generation?

4. What is the voltage of the electricity supply in the classroom?

Part V Task Implementation

Leading Questions

1. How many types of voltages are introduced in the dialogue?

2. Promote the core socialist values.

3. What is the voltage of the lighting electricity?

4. What does the reading recorded by the electricity meter indicate?

Part VI Task to Optimization

Leading Questions

1. What are the contents of electricity consumption indicators?

2. What happens when electricity comes into contact with water?

Module Eleven

Project Construction Phase（Ⅰ）

Task 1　Task Description
Task 2　Oral Practice
Task 3　Analysis of Civil Construction Knowledge

Task 1 Task Description

Part I Task Description

Understand the pre-bidding preparations required by the investor; understand the differences between open bidding and invitation for the tender; ensure that the bidding process complies with the provisions of the Tendering and Bidding Law of the People's Republic of China; and familiarize yourself with the duration of the public bidding period.

Part II Learning Goals

1. Master common words in oral English.
2. Understand common grammar used in oral English.
3. Acquire knowledge about the bidding process for construction projects.
4. Gain expertise in the execution of the bidding process.
5. Master the importance of abiding by the law in the bidding process.
6. Understand the role of drawings in the bidding process.

Part III Ideological and Political Points

1. Respect the integrity of the bidding process and ensure fair competition.
2. Strictly adhere to the laws and regulations governing the bidding process.
3. Embrace confidence, self-improvement, integrity, innovation, and determination as key characteristics for progress.
4. Utilize modesty, prudence, hard work, and courage to struggle.
5. Work diligently and honestly.
6. Show respect for the design work produced by designers.
7. The people are the foundation of the country, and the country is made up of its people.
8. Foster prosperity and development of cultural undertakings and industries.

Task 2 Oral Practice

Bidding Distribution

Diana—Investor
Jodie—Design Manager
Shrek—Project Manager

对话音频

Diana: Hello Shrek, can you please ask Jodie to prepare the complete drawings of Project A for bidding?

Shrek: Sure, when will you need them?

Diana: By Wednesday at the latest.

Shrek: I will call him immediately.

Jodie: Hi, Shrek, how are you recently?

Shrek: I am fine. Thank you.

Jodie: What is up?

Shrek: Could you print five sets of Project A-drawings for us? The bidding starts soon.

Jodie: Got it. I will personally deliver them to you on Tuesday.

Shrek: Thank you!

Wednesday morning

Diana: Shrek, the bidding is coming. Let us have an internal discussion about the drawings first.

Shrek: I have already informed everyone to gather in the meeting room for the discussion .

Two hours later

Diana: Alright, now tell me the outcome. Will it be an open bidding or an invited tendering?

Shrek: If it is an open bidding, we will need 5 days for the announcement. Otherwise, we need to select at least three companies for the invited tendering.

Diana: Maybe 5 days is too long. Let us start looking for qualified companies immediately.

Shrek: So you mean we should go with the invited tendering?

Diana: Yes, go ahead and proceed with it. Make sure to strictly follow the Tendering and Bidding Law of the People's Republic of China. Our goal is to "build a socialist rule of law system with Chinese characteristics and a socialist rule of law country, focusing on ensuring and promoting social fairness and justice."

Task 3 Analysis of Civil Construction Knowledge

Part Ⅰ Related Knowledge Links

Bidding	Open bidding	Invited tendering	the Tendering and Bidding Law of the People's Republic of China

Part Ⅱ Task Group

Class		Set No.		Tutor	
Group Leader		Student ID			
Team Members	Name	Student ID	Name	Student ID	
Division of Tasks					

Part Ⅲ Task Analysis

1. Is the bidding process carried out independently by the developers themselves?

2. Is the bidding process carried out independently by the Construction Company itself?

3. What are the differences between an open bidding and an invited tendering?

4. Why is it necessary to print the drawings before the bidding?

5. What is the purpose of conducting an internal discussion about the drawings?

6. Can the Design Company be responsible for the bidding process?

Part IV Access to Information

Leading Questions

1. Why is it not possible for the Construction Company to bid independently?

2. What is the relationship between the investor and the Tendering Company?

3. Can anyone participate in the bidding process?

4. How should problems with the drawings be addressed?

Part V Task Implementation

Leading Questions

1. What are the advantages of conducting the open bidding?

2. What are the advantages of using the invited tendering?

3. Do design companies provide their services to investors for free?

4. Why is there a public notice period for the public bidding?

Part VI Task to Optimization

Leading Questions

1. What are the restrictions on the bidding process?

2. What preparations should be made for the tender process?

Module Twelve

Project Construction Phase (Ⅱ)

Task 1 Task Description

Part I Task Description

Know which company should issue the bidding invitation letter; know that the tenderer needs to submit an application for pre-qualification; be familiar with the pre-tender period after the bidding documents are issued; understand the requirements for the number of bidding experts at the bid opening site; and know when and where to notify the winning bid result.

Part II Learning Goals

1. Master common words in oral English.
2. Understand the common grammar used in oral English.
3. Understand the bid opening process.
4. Understand the importance of submitting bids on time.
5. Understand the precautions in the bidding process.
6. Recognize the importance of responding to questions on the bidding site.

Part III Ideological and Political Points

1. Adhere to the geological survey findings during design and construction.
2. Promote fairness and respect towards competitors.
3. Marxism is the guiding ideology for building and revitalizing the Party and the country.
4. The bill of quantities must be accurate and correct.
5. The information of the Tendering Company must be protected before the bid opening.
6. Widely promote the core socialist values.
7. Approach science with a scientific attitude.

Task 2 Oral Practice

Tendering Process

对话音频

Diana—Investor
Alex—Vice General Manager
Shrek—Project Manager
Divian—Tenderer
Chris—Tenderer
Vivid—Tenderer
Tallmi—Manager of a Project Management Company

Chris's Company

Chris: Guys, today we received invitation to bid from Shrek. We don't have much time left, so let us hurry and get ready.

Employee: OK, boss!

Divian's Company

Divian: I'm so excited! We received the invitation to bid. I always had doubts about the strength of our company. Let us carefully check the bid registration information.

The Secretary: We will put in great efforts. Trust us.

Vivid's Company

Vivid: Let us thoroughly study the bidding project information. Our competitors are strong, so we must be extra cautious. There should be no mistakes in the tender documents.

The Secretary: We have completed and submitted the application for pre-qualification.

After 2 days

Tallmi: Diana,I have received applications for pre-qualification of three companies. I will review and analyse right now.

Diana: Thank you for all your hard work.

Next day

Tallmi: Diana, we have ranked the three companies. Please check and determine the qualified bidders.

Diana: Let me take a look. OK! Go with your choice. Issue the notice of pre-qualification.

Tallmi: OK!

Shrek: I will organize them to survey the site.

Next week

Tallmi: Diana,I am preparing the bidding documents.

Diana: Send them to the invited tenderers once you're finished.

Tallmi: I will send them out tomorrow. The pre-tender period is 20 days.

Vivid: Do not forget the receipt of bidding documents .

The Secretary: Let us take action right now.

Divian: While preparing the tender documents, let us also optimize the drawings.

The Secretary: OK, I will arrange that.

Chris: Collect all the relevant information before preparing the tender documents.

Employee: That is already in progress, as you instructed.

Tallmi: Shrek, could you come to our company this afternoon to answer tenders' questions face to face?

Shrek: Did you collect their written questions?

Tallmi: Yes, we will clarify and modify the bidding documents accordingly.

Divian: Is everything clear now?

The Secretary: Yup, we are taking care of the bid security.

Chris: Tallmi, these are our tender documents.

Divian: These are ours.

Vivid: And mine.

Tallmi: OK, my secretary will write down the date and time of receipt.

Tallmi: Miss Chen, send back the overdue tender document. Diana, we received the documents from three companies on time.

Diana: Start preparing the list of materials and the pre-tender estimated price.

Tallmi: We will organize the bid opening.

Diana: Have you found the bidding evaluation experts?

Tallmi: Yes, we have.

Bid opening site

Tallmi: Experts, there are three tender documents in front of you. Please score them based on fairness.

Several hours of bid evaluation later

Expert: Let them answer questions one by one.

Tallmi: We will rank the companies based on the scores given by the experts. Now, we recommend Divian's Company, which received the highest scores, as the winning bidder. The public announcement will be made within three working days.

Diana: Inform all the tenderers about the result.

Shrek: I will inform Divian to sign contract agreement. And we will return the bid security to all of them.

Diana: Okay, even though the bidding work is over, "everything is interconnected and interdependent." Let us do a good job of continuing with the next tasks.

Task 3　Analysis of Civil Construction Knowledge

Part I　Related Knowledge Links

Pre-qualification	Geological investigation	Tender receipt	Bid security

Part II　Task Group

<table>
<tr><td>Class</td><td></td><td>Set No.</td><td></td><td>Tutor</td><td></td></tr>
<tr><td>Group Leader</td><td></td><td>Student ID</td><td></td><td></td><td></td></tr>
<tr><td>Team Members</td><td colspan="5"><table>
<tr><td>Name</td><td>Student ID</td><td>Name</td><td>Student ID</td></tr>
<tr><td></td><td></td><td></td><td></td></tr>
<tr><td></td><td></td><td></td><td></td></tr>
</table></td></tr>
<tr><td>Division of Tasks</td><td colspan="5"></td></tr>
</table>

Part III　Task Analysis

1. Why should bidding companies go through the pre-qualification process?
2. Why is it necessary to conduct a site survey in advance?
3. Who is responsible for conducting the site survey?
4. Why do we need proof of receipt for tender documents during the bidding process?
5. Why is it necessary to have a review expert during the bidding process?
6. Why is it important to specify the tender price when bidding?

Part IV　Access to Information

1. Why is bid security required?

____

____

2. Why is it important to publicize the winning result?

3. How does the geological survey results affect the project?

4. Can bidding companies discuss with each other?

Part V Task Implementation

Leading Questions

1. How should a bidding company be selected?

2. Should the Design Company adhere to the findings of the geological survey?

3. Who is responsible for preparing the bill of quantities?

4. How should the situation be handled if the project price is lower than the tender price?

Part VI Task Optimization

Leading Questions

1. What are the risks of not following the geological survey results?

2. Why should the bid bond be returned to a company that did not win the bid?

Module Thirteen

Project Construction Phase (Ⅲ)

Task 1 Task Description
Task 2 Oral Practice
Task 3 Analysis of Civil Construction Knowledge

Task 1 Task Description

Part I Task Description

Understand the necessary preparations before the construction begins; ensure that the drawings are complete and meet the construction requirements during the construction phase; learn about the specific work content related to "three supplies and one leveling" ; familiarize yourself with local data on natural conditions and technical experience; take measures to protect the ecological environment during construction; and complete the installation of temporary facilities and public facilities prior to the start of construction.

Part II Learning Goals

1. Master common words in oral English.
2. Understand the common grammar used in oral English.
3. Master the pre-construction preparations.
4. Understand the concept of "three supplies and one leveling".
5. Recognize that coordination is crucial for producing quality work.
6. Understand the importance of all procedures in the construction process.

Part III Ideological and Political Points

1. Take the work of "three supplies and one leveling" before construction seriously.

2. Carefully examine the local natural conditions.

3. Empty talk is detrimental to the nation, while practical work leads to its prosperity.

4. A strong youth leads to a strong country.

5. Supervising engineers should diligently supervise the work and serve the Construction Company well. Firmly follow the leadership of our Party.

6. Temporary facilities should be constructed with proper safety measures.

7. The people of all nationalities should unite, share the same fate, and have a harmonious relationship.

Task 2 Oral Practice

Construction Preparation

Helen—Construction Project Manager
Kevin—Civil Engineer
Mary—Equipment Engineer

对话音频

Helen: Please carefully review the design drawings and familiarize yourselves with the relevant materials and design data.

Kevin: I have already reviewed the drawings roughly. The contents are complete, but some details are missing.

Helen: Make notes and provide feedback to the Design Company during the technical disclosure.

Mary: Some machine room drawings have not been provided, such as power transformation and distribution room, heat exchange station, and other utility rooms.

Helen: Those will be provided by the Electric Power Design Institute and the Thermal Company. Do not worry, the Construction Company will coordinate with them separately.

Mary: OK!

Helen: Kevin, go to the construction site and gather information about the current situation. Collect local data on natural conditions and technical experience. We adhere to the concept that lucid waters and lush mountains are invaluable assets and emphasize ecological and environmental protection in all aspects, regions, and processes.

Kevin: Already done.

Helen: How is the preparation?

Kevin: We have completed the "three supplies and one leveling" and the site lofting. They have been confirmed by the supervising engineer.

Helen: You both need to complete the construction of temporary facilities and public facilities as soon as possible.

Mary: Got it.

Helen: Implement all the material requirements.

Kevin: They are being implemented one by one.

Helen: Do not make any mistakes in the preparation of the construction team.

Kevin: The labor team has been deployed and organized.

Helen: Issue the operation plan and construction assignments to them.

Kevin: I will get ready immediately.

Helen: Let us check everything together and ensure that every detail is done well to guarantee a smooth start of construction.

Mary: Okay!

Kevin: Got it.

Task 3 Analysis of Civil Construction Knowledge

Part Ⅰ Related Knowledge Links

Three supplies and one leveling	Building node	Public facility	Temporary facility

Part Ⅱ Task Group

<table>
<tr><td>Class</td><td></td><td>Set No.</td><td></td><td>Tutor</td><td></td></tr>
<tr><td>Group Leader</td><td></td><td>Student ID</td><td></td><td></td><td></td></tr>
<tr><td>Team Members</td><td colspan="5"><table><tr><td>Name</td><td>Student ID</td><td>Name</td><td>Student ID</td></tr><tr><td></td><td></td><td></td><td></td></tr><tr><td></td><td></td><td></td><td></td></tr></table></td></tr>
<tr><td>Division of Tasks</td><td colspan="5"></td></tr>
</table>

Part Ⅲ Task Analysis

1. Why is "three supplies and one leveling" required before construction?
2. What should be done if the drawing contents are incomplete?
3. Does weather condition impact the construction schedule?
4. Why is it important for the Construction Company to create a construction schedule?
5. Does changes in construction drawings lead to increased costs?
6. Does an unfinished drawing affect the construction quotation?

Part Ⅳ Access to Information

Leading Questions

1. Can incomplete drawings be used for construction?

2. Can incomplete drawings be used for construction ?

3. What authority does the supervising engineer have in the project?

4. Is the supervising engineer a member of the investor?

Part V Task Implementation

Leading Questions

1. What does "three supplies and one leveling" refer to?

2. How does the local economic condition affect the project?

3. What is the purpose of gathering site formation?

4. What materials need to be brought in before construction?

Part VI Task Optimization

Leading Questions

1. How does the natural condition impact the project?

2. What contents are included in the construction assignment?

Module Fourteen

Project Construction Phase (Ⅳ)

Task 1 Task Description

Part I Task Description

Understand the importance of creating a construction schedule; take into account weather and other unforeseen circumstances when developing the construction schedule; the construction schedule should be developed collaboratively with input from various professions; understand the concept and purpose of a Gantt chart; and know which company should the schedule be submitted.

Part II Learning Goals

1. Master common words in oral English.
2. Understand the common grammar used in oral English.
3. Master the content of the project construction schedule.
4. Understand the importance of dividing the construction schedule by division.
5. Master the correct construction process.
6. Understand how to establish a reasonable construction schedule.

Part III Ideological and Political Points

1. Strictly adhere to the agreed contractual period of time and develop a sense of time quality.Conduct yourself with integrity and cleanliness.
2. Prepare countermeasures in advance for unforeseen circumstances.
3. Show unwavering determination and commitment.
4. Train students to apply their knowledge to solve practical problems.
5. Train students to draw logical conclusions through practice.
6. Realize the Chinese Dream of national rejuvenation.
7. Show loyalty to the Party and dedication to serving the country and the people.

Task 2 Oral Practice

Construction Schedule

Helen—Construction Project Manager
Kevin—Civil Engineer
Mary—Equipment Engineer

对话音频

Helen: According to the contract, we are expected to complete the entire project within one and a half years. Now, let us discuss the construction schedule. "During the construction process, we should implement a comprehensive energy-saving strategy and promote the conservation and intensive utilization of various resources."

Kevin: After the meeting, I will immediately create a general schedule for the project and ask them to prepare the relevant materials.

Mary: After the civil construction time-frame is made, I will make the installation schedule.

Helen: The construction time-frame is tight, but we should also take into account potential delays caused by weather conditions. So, please ensure that we make the most of our time.

Kevin: I will take climatic factors into full consideration when plan the monthly progress of the project.

Helen: This is particularly important for outdoor construction.

Kevin: OK!

Mary: Please consider the underground pipes when creating the progress chart for outdoor road construction.

Kevin: We can discuss this part of the schedule together.

Helen: All the relevant information about the progress must be well prepared.

Kevin: Got it!

Helen: The objectives for duration management should be clearly marked.

Kevin: Yes, OK!

Mary: Once I review your overall plan, we can confirm the entry time for the installation section at milestone nodes.

Helen: Let us use the level 1 network plan for the overall schedule. For the monthly progress plan, we can use a Gantt chart.

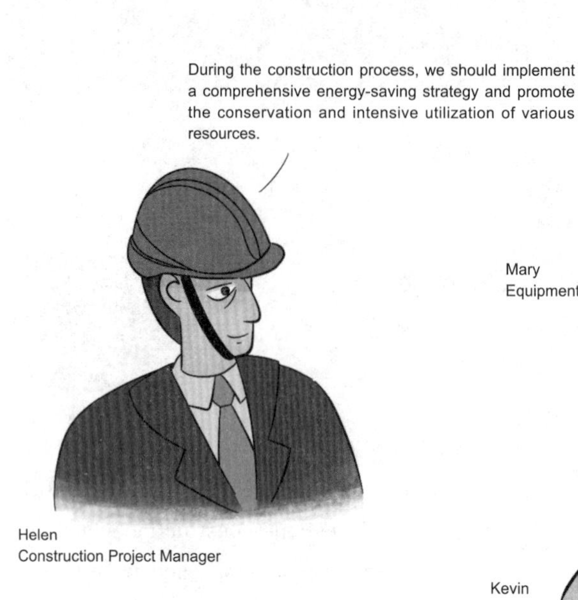

Kevin: Got it!

Helen: In case any unforeseen circumstances cause the project delays, any extension of time needs to be justifiable and convincingly explained.

Kevin: I will.

Helen: The progress plan shall be submitted to the Construction Company and the Supervision Company at least seven days in advance.

Kevin: I see. I will organize the project accordingly and strive to complete it in the shortest time-frame possible.

Helen: Let us prepare separately for now.

Mary: OK!

Task 3　Analysis of Civil Construction Knowledge

Part Ⅰ　Related Knowledge Links

Construction schedule	Gantt chart	Network plan	Supervision Company

Part Ⅱ　Task Group

<table>
<tr><td>Class</td><td></td><td>Set No.</td><td></td><td>Tutor</td><td></td></tr>
<tr><td>Group Leader</td><td></td><td>Student ID</td><td></td><td></td><td></td></tr>
<tr><td rowspan="3">Team Members</td><td>Name</td><td>Student ID</td><td>Name</td><td colspan="2">Student ID</td></tr>
<tr><td></td><td></td><td></td><td colspan="2"></td></tr>
<tr><td></td><td></td><td></td><td colspan="2"></td></tr>
<tr><td>Division of Tasks</td><td colspan="5"></td></tr>
</table>

Part Ⅲ　Task Analysis

1. Why is it necessary to divide the construction schedule into overall schedule and monthly schedule?

2. Why should the construction schedule be submitted to the Supervision Company for approval seven days in advance?

3. How do you plan and schedule the project duration?

4. Will longer construction periods result in higher profits?

5. Does making more design changes lead to increased profits?

Part Ⅳ　Access to Information

Leading Questions

1. What are force unforeseen circumstances and how many factors do you know about them?

2. Can the construction schedule restrict the construction period completely?

3. Why can not a design firm handle the project management for the same project?

4. Do you believe that more design changes always improve a project?

Part V Task Implementation

Leading Questions

1. What would happen if there were no construction schedule?

2. Why can not the Construction Company just make the construction schedule?

3. Who is responsible for any cost increase resulting from design changes?

4. Can the project be submitted without going through an acceptance process?

Part VI Task Optimization

Leading Questions

1. Does the construction schedule need to be reviewed by the Construction Company?

2. Please describe the sequence of construction for your current classroom project.

Module Fifteen

Project Construction Phase（V）

Task 1 Task Description
Task 2 Oral Practice
Task 3 Analysis of Civil Construction Knowledge

Task 1 Task Description

Part I Task Description

Understand the importance of surveying and staking before foundation construction. Formwork support is necessary during the excavation of foundation pits. Workers should undergo safety training before carrying out reinforcement binding. Familiarize yourself with the avoidance of pouring concrete during rainy weather. Ensure accurate calculation of backfill earthwork volume.

Part II Learning Goals

1. Master common words in oral English.
2. Understand the common grammar used in oral English.
3. Understand the importance of building foundations for the overall structure.
4. Understand the purpose of accurate pay-off in construction.
5. Understand the reasons why concrete should not come into contact with water.
6. Know when it is necessary to purchase backfill materials.

Part III Ideological and Political Points

1. Train students to prioritize safety awareness at all times.
2. Accurate positioning in pay-off is crucial for the entire building.
3. Dedicate ourselves to the noble cause of peace and human development.
4. Stay true to our original aspirations and the founding mission.
5. Ensure precise tying of reinforcement bars.
6. Do not underestimate the importance of addressing even small acts of evil.
7. Maintain a strategic composure and carry forward the spirit of struggle.

Task 2 Oral Practice

Foundation Construction

Helen—Construction Project Manager
Kevin—Civil Engineer

对话音频

Helen: Has the positioning and setting out of foundation column been completed?

Kevin: Yes, and I have scheduled the technical disclosure for this afternoon.

Helen: Is the subcontractor of woodworking formwork on-site?

Kevin: They arrived at the site yesterday.

Helen: The workers responsible for reinforcing the foundation platform must receive safety training in advance. "Adhere to the principle of putting people first and prioritizing safety."

Kevin: Joe will be in charge for it.

Helen: The positioning of electric welding for the reinforcement of the foundation platform and columns must continually improve.

Kevin: Clear!

Helen: Tell Cute that the embedding of water and electricity must be accurate to avoid rework.

Kevin: I will review the drawings with her tomorrow.

Helen: By the way, there may be rain in the coming days. Try to avoid pouring concrete during rainy weather to protect it from getting wet.

Kevin: Rest assured.

Helen: When removing the formwork, the subcontractor must clean it thoroughly to ensure it doesn't affect the acceptance of the foundation.

Kevin: I will supervise that carefully.

Helen: Make sure to carefully calculate the volume of backfill earthwork that needs to be purchased and strive to save costs within the specified requirements.

Kevin: Clear!

Helen: I need to leave right now. You can proceed with other tasks.

Kevin: OK!

Task 3　Analysis of Civil Construction Knowledge

Part Ⅰ　Related Knowledge Links

Pile foundation platform	backfilling	Formwork support	Subcontractor

Part Ⅱ　Task Group

<table>
<tr><td>Class</td><td></td><td>Set No.</td><td></td><td>Tutor</td><td></td></tr>
<tr><td>Group Leader</td><td></td><td>Student ID</td><td></td><td></td><td></td></tr>
<tr><td>Team Members</td><td colspan="5"><table><tr><td>Name</td><td>Student ID</td><td>Name</td><td>Student ID</td></tr><tr><td></td><td></td><td></td><td></td></tr><tr><td></td><td></td><td></td><td></td></tr></table></td></tr>
<tr><td>Division of Tasks</td><td colspan="5"></td></tr>
</table>

Part Ⅲ　Task Analysis

1. Why do we use formwork for cast-in-situ concrete construction?
2. What is the relationship between subcontractors and contractors?
3. Why is safety training necessary for construction personnel before starting work?
4. How to ensure concrete moisture resistance?
5. What are the requirements for backfill materials?
6. What conditions must be met for removing supporting formwork?

Part Ⅳ　Access to Information

Leading Questions

1. Who is responsible for the foundation pay-off?

2. Can the construction unit handle formwork support work independently?

3. Is the amount of backfill materials equal to the excavated earth material?

4. Are there any special requirements for backfill?

Part V Task Implementation

Leading Questions

1. Can subcontractors communicate directly with the investor?

2. How many forms of concrete foundation?

3. Can the subcontractor communicate directly with the investor?

4. Is it necessary to refer to the drawing for reinforcement bar bending and binding?

Part VI Task Optimization

Leading Questions

1. What contents are included in the construction assignment?

2. What is the minimum cover depth?

Module Sixteen

Project Construction Phase（Ⅵ）

Task 1　Task Description
Task 2　Oral Practice
Task 3　Analysis of Civil Construction Knowledge

Task 1 Task Description

Part I Task Description

Rectify timely problems during construction; analyze correctly the cause of accidents. The Construction Company should propose a rectification approach. Know the importance of scientific construction.

Part II Learning Goals

1. Master common words in oral English.Master the process of identifying and reporting construction problems to investors during construction supervision.

2. Understand the common grammar used in oral English.

3. Develop the ability to promptly identify and address construction problems during the construction process.

4. Understand the importance of construction quality as a matter of life and death.

5. Understand that construction quality can be guaranteed when everyone has a sense of responsibility and mission.

Part III Ideological and Political Points

1. Help students understand the significance of science and technology in national rejuvenation.Encourage students to tackle difficult tasks and take calculated risks.

2. Foster students' habits of independent learning and positive thinking.

3. Encourage students to face new contradictions and challenges with courage.

4. Guide students to develop good professional ethics in all aspects.

5. Enhance students' understanding of the architectural spirit in the new era.

6. Our motherland now boasts bluer skies, greener mountains, and clearer waters.

Task 2 Oral Practice

Construction Rectification

Shrek—Project Manager
Bob—Supervising Engineer
Helen—Construction Project Manager

对话音频

Shrek: Now, let the Supervising Company report the construction problems found on site yesterday.

Bob: Based on the site inspection, there are early cracking issues with the pouring plate in workshop 2. There are roughly two cases: one is general cracking in the center of the pouring plate, parallel to the trench at the plate's edge; the other is early cracking caused by improper treatment of the auxiliary drainage ditch connection.

Shrek: Now, let the Construction Company analyze the cause of the cracking.

Helen: In the preliminary analysis, three factors were identified:

1. The floor slab is constrained by the trench wall on both sides, leading to shrinkage cracking.

2. Poor concrete performance leads to prolonged bleeding and inadequate condensation.

3. Environmental factor.

Shrek: How can we solve it?

Helen: The technical staff is currently working on a response plan and will submit it in written form for review by the Supervision Company.

Bob: The repair plan and materials for the initial crack should also be included in the report.

Helen: All right.

Shrek: Are there any other construction problems?

Bob: The approach between the structure and the building floor has also been documented.

Helen: Got it.

Shrek: Take the necessary time to rectify these issues to avoid impacting the construction schedule.

Helen: We will speed up the rectification work.

Shrek: Anything else?

Bob: Nothing more!

Shrek: I hope we can build scientifically and address potential risks proactively, advancing the construction of a safer China to a higher level. Over!

Task 3 Analysis of Civil Construction Knowledge

Part Ⅰ Related Knowledge Links

Slab cracked factors	Crack repair	Construction organization scheme	Construction rectification

Part Ⅱ Task Group

Class		Set No.		Tutor	
Group Leader		Student ID			
Team Members	Name	Student ID	Name	Student ID	
Division of Tasks					

Part Ⅲ Task Analysis

1. Can all construction problems be resolved through meetings?

2. What should the Supervision Company do if the Construction Company does not address the construction problem identified by the Supervision Company?

3. When you encounter a problem, do you first blame yourself or others?

4. Should everyone have a sense of responsibility?

5. How can you develop yourself into a responsible individual?

Part IV Access to Information

Leading Questions

1. What are the consequences if construction problems are not identified in a timely manner?

____

____

2. How should unqualified raw materials found during construction be handled?

____

____

3. Is it necessary to create a construction plan before starting the construction project?

____

____

4. How can construction problems caused by environmental factors be avoided?

____

____

Part V Task Implementation

Leading Questions

1. Why can environmental factors contribute to construction problems?

____

____

2. Does the skill level of construction personnel play a role in construction problems?

____

____

3. Does the investor solely rely on the Supervision Company during the construction process?

____

____

4. What professional qualities should the Supervision Company have to avoid construction problems?

____

____

Part VI Task Optimization

Leading Questions

1. Provide examples of poorly executed construction projects you are aware of and analyze the causes of the accidents.

____

____

2. What professional qualities should a construction company possess to avoid construction problems?

____

____

Module Seventeen

Project Construction Phase（Ⅶ）

Task 1　Task Description
Task 2　Oral Practice
Task 3　Analysis of Civil Construction Knowledge

Task 1 Task Description

Part I Task Description

Familiarize with precautions during pipeline installation; be able to independently analyze the causes of pipeline leakage; understand the significance of using qualified products in engineering projects; conduct pipeline pressure testing in accordance with specifications and design requirements; and understand the significance of paying attention to details.

Part II Learning Goals

1. Master common words in oral English.
2. Understand the common grammar used in oral English.
3. Understand the installation requirements for water supply and drainage pipes.
4. Master the nominal pressure requirements of water supply and drainage pipes.
5. Understand that success or failure depends on attention to details.
6. Recognize that the quality of installation determines the safety of life and property.

Part III Ideological and Political Points

1. Understand the negative impacts on life and production caused by excessive pressure in pipelines.
2. Emphasize the importance of scientific and proper pipe installation.
3. Learn from history to gain insight into reality.
4. Advocate an inclusive and global approach.
5. Detail determines success or failure.
6. Quality is the basis of a hundred-year plan.
7. Uphold the harmonious coexistence between humans and nature.
8. Promote sustainable development.

Task 2 Oral Practice

Piping Installation Engineering

Mary—Equipment Engineer
Bob—Supervising Engineer
Jony—Piping Worker

对话音频

Bob: Ah, Mary, Hurry up! I hope you would not be long. We have a big trouble here.

Mary: Well? What happened?

Bob: Just as what I said, it is a crisis. You see, it is a prophecy. There is a severe pipe leakage in workshop 2, and the wall is seriously damaged.

Mary: Oh, dear! All right, I will meet you on-site in 15 minutes.

Bob: Wait for you there. Take the worker with you so we can easily conduct field maintenance.

Mary: Got it. I will call him immediately.

Jony: Hello, Mary, I am busy installing radiator. Get straight to the point, please.

Mary: There is a pipe leakage in workshop 2. Come here right now.

Jony: Really? What is wrong? As far as I remember, it was well-installed. I can not leave now. Can you give me a second? Please!

Mary: Leave the radiator there. It does not matter. Bob is also waiting for us there. Hurry! there is no time to lose!

Jony: OK, You are the boss. Be there or be square!

Bob (in angry): Did you check it? Was it properly tested? What was the pressure? And is the raw material qualified?

Jony: Cool down, Bob. Let me find out what is going on.

Bob: Hurry! I do not want to see such a mess one hour from now!

Jony: Anyway, I am sorry for what has happened.

Mary: Hi, Jony, where is the leakage?

Bob: Jony, we have a 100-year plan and a focus on quality. Can you take your job seriously?

Jony: Bob, honestly, the pipe quality is terrible.

Bob: What do you mean? Is it nonconforming product?

Jony: Sorry, I am not sure, you can conduct a survey...

Bob: Jony, can you give me an explanation?

Mary: Jony, what was the test pressure?

Jony: Let me see... Oh, it was 1.6MPa.

Mary: Oh my god! Overpressure! This is a domestic water supply pipe. The pipe's working pressure is only 0.60MPa. Testing it up to 1.10MPa is sufficient! Stop testing immediately!

Bob: Mary, you should supervise every detail throughout the entire installation process. Avoid similar crises, okay?

Mary: Sorry!

Jony: Sorry Bob! I will solve the problem immediately.

Bob: The devil is in the details! I hope we can prioritize quality assurance engineering. China speed and China quality come first! Let us start with ourselves. Success can only be achieved by capable people.

Task 3 Analysis of Civil Construction Knowledge

Part Ⅰ Related Knowledge Links

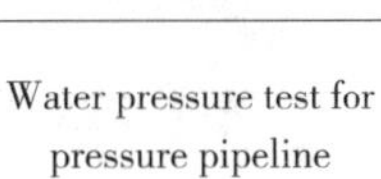 Water pressure test for pressure pipeline	Pneumatic water supply	Regulations on quality control of construction projects	Secondary water supply

Part Ⅱ Task Group

<table>
<tr><td>Class</td><td></td><td>Set No.</td><td></td><td>Tutor</td><td></td></tr>
<tr><td>Group Leader</td><td></td><td>Student ID</td><td></td><td></td><td></td></tr>
<tr><td rowspan="3">Team Members</td><td>Name</td><td>Student ID</td><td>Name</td><td colspan="2">Student ID</td></tr>
<tr><td></td><td></td><td></td><td colspan="2"></td></tr>
<tr><td></td><td></td><td></td><td colspan="2"></td></tr>
<tr><td>Division of Tasks</td><td colspan="5"></td></tr>
</table>

Part Ⅲ Task Analysis

1. How to determine the diameter of water supply and drainage pipe?
2. Why is there pressure in the water supply line?
3. What are the requirements for water supply pipe materials?
4. Will the cost of pipe disassembly and re-installation increase during construction?
5. What are the differences between nominal stress and working stress?
6. What are the consequences of using nonconforming products?

Part Ⅳ Access to Information

Leading Questions

1. What are the causes of pipe over-pressure during installation?

2. Can the same materials be used for water supply and drainage pipes?

3. How do you choose the materials for water supply pipes?

4. Is the flow rate in the pipeline a fixed value?

Part V Task Implementation

Leading Questions

1. Does the water supply and drainage pipe have any impact on water quality?

2. Does the diameter of the water supply and drainage pipe affect the water pressure?

3. How do you determine the flow rate of water in a pipe?

4. How do you fix pipeline leakage?

Part VI Task Optimization

Leading Questions

1. How can water in pipes be prevented from being polluted?

2. Who will supervise the quality and progress of the entire installation process?

Module Eighteen

Project Completion Phase

Task 1 Task Description
Task 2 Oral Practice
Task 3 Analysis of Civil Construction Knowledge

Task 1 Task Description

Part I Task Description

Understand that a completion report should be issued after construction acceptance. The completion report should be signed and confirmed by multiple parties before it can be considered valid. The quality of each individual project must meet the requirements for acceptance. During the acceptance process, the Construction Company should conduct self-evaluation as well as receive evaluation from the Supervision Company.

Part II Learning Goals

1. Master common words in oral English.
2. Understand the common grammar used in oral English.
3. Master the contents of the completion report.
4. Understand which company should issue the completion report.
5. Master the whole process of a project from preparation to completion.
6. Know that all departments are required to participate in the project completion and acceptance.

Part III Ideological and Political Points

1. Prioritize quality, ensuring that all requirements are met.
2. Respect the acceptance opinions of each inspection group and address any mistakes.
3. Promote green and low-carbon production and lifestyle.
4. Emphasize individual responsibility.
5. Embrace the spirit of craftsmanship and do your best in your work.
6. Develop a professional quality of caring for others, solidarity and friendship.
7. Never be swayed by wealth, poverty, or force.

Task 2 Oral Practice

Completion Report

Diana—Investor
Shrek—Project Manager
Helen—Construction Project Manager
Jodie—Design Manager
Bob—Supervising Engineer

对话音频

Diana: The next step is to prepare the completion report.

Shrek: I have prepared the construction permits and drawing review comments.

Jodie: The general situation of the project can be extracted from the construction drawing description.

Diana: Are all the procedure documents prepared for the entire project implementation process?

Shrek: Everything is complete except the engineering quality acceptance report.

Helen: The project quality acceptance report also needs the signature of the designer and the seal of the Design Company.

Jodie: Just go to our company and find the appropriate designer.

Helen: A copy of the installation acceptance still needs to be stamped by the Supervision Company.

Bob: After the meeting, the seal was arranged immediately. The person in charge of the seals went back to the company yesterday.

Helen: OK!

Diana: The quality of the foundation, subgrade, and main structure must meet the mandatory standards for engineering construction, as well as the design and contract requirements. It should ensure that the information is complete and qualified.

Shrek: Got it. The quality of the roof engineering should also conform to engineering standards and design, without any leakage, meet the functional requirements, and ensure that the information is complete and qualified.

Bob: So does the quality of the installation project. It should also conform to engineering standards and designs, and the functions must be tested to meet the requirements. The information should be complete and qualified.

Diana: OK!

Bob: There is no problem with the quality inspection of the fire protection project, but the acceptance report can only be retrieved tomorrow.

Helen: Regarding the overall project quality our self-evaluation is qualified.

Bob: The evaluation given by the Supervision Company for the overall project is qualified.

Shrek: Next, we will wait for the comprehensive acceptance of the project completion.

Diana: OK, each of you should finish the remaining tasks, and let us look forward to a successful project. Let us adhere to serving the people and socialism, guided by the core socialist values, and promote advanced socialist culture.

Task 3　Analysis of Civil Construction Knowledge

Part Ⅰ　Related Knowledge Links

Completion report	Construction permit	Main structure	Fire protection acceptance

Part Ⅱ　Task Group

<table>
<tr><td>Class</td><td></td><td>Set No.</td><td></td><td>Tutor</td><td></td></tr>
<tr><td>Group Leader</td><td></td><td>Student ID</td><td></td><td></td><td></td></tr>
<tr><td>Team Members</td><td colspan="5"><table><tr><td>Name</td><td>Student ID</td><td>Name</td><td>Student ID</td></tr><tr><td></td><td></td><td></td><td></td></tr><tr><td></td><td></td><td></td><td></td></tr></table></td></tr>
<tr><td>Division of Tasks</td><td colspan="5"></td></tr>
</table>

Part Ⅲ　Task Analysis

1. Which company will take charge of the completion quality inspection?

2. Does the supervision work issue the acceptance report during the completion quality acceptance?

3. Should the Design Company also participate in the completion quality inspection?

4. Who is responsible for completing the quality inspection of fire control project?

5. Why must the quality of foundation and main structure meet the mandatory standards of engineering construction?

6. Are there any safety risks during operation of the installation project?

Part IV Access to Information

Leading Questions

1. Is the Supervision Company responsible for quality problems in the project?

____

____

2. Is the Investor Company responsible for quality problems?

____

____

3. Does the Investor Company participate in the project acceptance?

____

____

4. Can the accepted project be put into operation immediately?

____

____

Part V Task Implementation

Leading Questions

1. How to solve quality problems identified in the project?

____

____

2. Who will bear the costs incurred during the rectification process if quality problems are found in the project?

____

____

3. If some parts of the project do not pass the completion acceptance, can they still be operated during the rectification period?

____

____

4. Who will be responsible for overseeing the project after completion?

____

____

Part VI Task Optimization

Leading Questions

1. Based on the content of this dialogue, summarize the information and create a completion report.

____

____

2. Now the project has been completed. Please share your overall thoughts and feelings.

____

____

Module Nineteen

Green Building（Ⅰ）

Task 1 Task Description
Task 2 Oral Practice
Task 3 Analysis of Civil Construction Knowledge

Task 1 Task Description

Part I Task Description

Familiarize yourself with the concept of Green Buildings; understand the profound meaning of the value of pristine environment; recognize the symbolic significance of the color green; and grasp the various aspects that Green Buildings encompass.

Part II Learning Goals

1. Master common words in oral English.
2. Understand the common grammar used in oral English.
3. Understand the importance of protecting nature from pollution.
4. Understand the true meaning of Green Buildings.
5. Master the impact of Green Buildings on people's life and work.
6. Know the health benefits of clean air in your living environment.

Part III Ideological and Political Points

1. Lucid water and lush mountains are invaluable assets.
2. Take care of every tree and grass in nature.
3. Each individual should strive to be a messenger in protecting nature and the environment.
4. Love nature and live in harmony with nature.

Task 2 Oral Practice

What is the Green Building

Kathy—Teacher
Polly—Student
Teddy—Student
Michael—Student
Linda—Student

对话音频

Kathy: What is a Green Building? Does anyone know?

Polly: It is a building with plants.

Teddy: The color of exterior wall is green.

Michael: Oxygenated building?

Linda: I have no clue.

Polly: Am I wrong? I don't think so.

Kathy: Let me explain it step by step.

Michael: Madam, would you like to explain first, does green mean a color?

Kathy: OK, you noticed the key point. Here is another question: what does green symbolize?

Teddy: Peace.

Michael: Youth.

Linda: Love.

Teddy: Hope.

Kathy: Yeap, you all are right. For the modern significance of Chinese culture, green stands for ecology and environmental protection.

Linda: Madam, it hits me, this green is similar to "Lucid waters and lush mountains are invaluable assets." said by our Chairman Xi Jinping.

Kathy: You are a smart cookie! The 20th National Congress of the Communist Party of China mentions that "we will deepen the prevention and control of environmental pollution, continue to fight well in the defense of blue skies, clear waters,and clean land, basically eliminate heavy polluting weather, and basically eliminate urban black and odorous water bodies."

Teddy: I have a question, madam.

Kathy: Please go ahead.

Teddy: Do we need to use clean water to build the building?

Michael: What is the relationship between green and building?

Polly: There should be green plants in the building, just like I mentioned earlier.

Kathy: Not exactly, Polly. OK! Hold on, let me drink some water (with a smile).

Linda: Take your time.

Kathy: Let us start with green plants.

Michael: You mean plants, Madam?

Kathy: As you all know, plants can purify the air, wherever we place them. They release oxygen and absorb carbon dioxide.

Teddy: Is this what rooftop gardens are designed for?

Kathy: The rooftop garden is not just about purifying the air. Let's imagine, where does rainwater from the roof go?

Linda: It flows from the roof to the ground through vertical rain pipe on the exterior wall.

Kathy: What color is the rainwater?

Polly: Almost black. It is very dirty.

Kathy: The plants on the roof can intercept rainwater with their roots, preventing pollution of the ground.

Michael: So it is about environmental protection?

Kathy: Great! Meanwhile the plants can absorb sunlight. It acts as insulation in the summer.

Linda: Got it.

Kathy: This is only a part of Green Building. We will continue in the next class. Take a break.

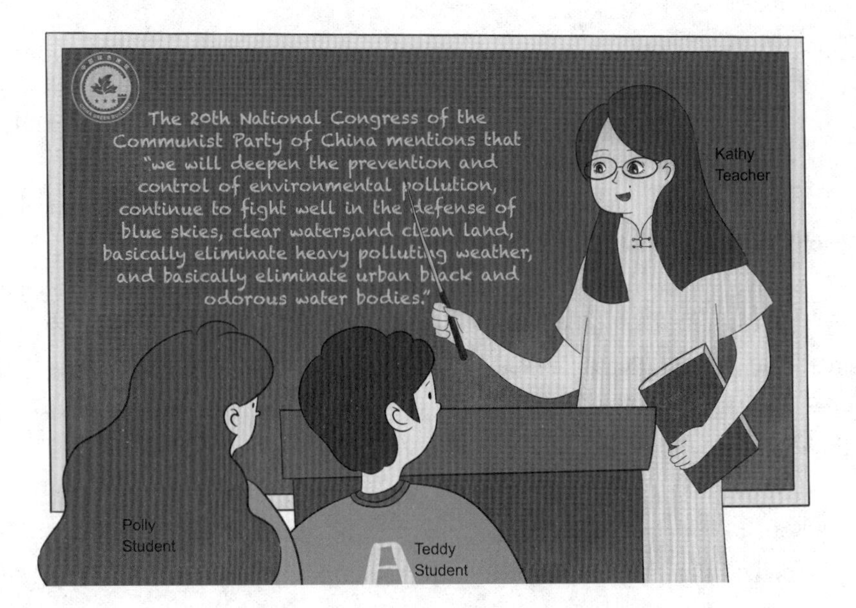

Task 3 Analysis of Civil Construction Knowledge

Part I Related Knowledge Links

Lucid waters and lush mountains are invaluable assets	Reclaimed water	Organized drainage

Part II Task Group

Class		Set No.		Tutor	
Group Leader		Student ID			
Team Members	Name	Student ID	Name	Student ID	
Division of Tasks					

Part III Task Analysis

1. Does "Green Building" refer to the color of the building being green?
2. How can a building be considered green?
3. Why should we promote Green Buildings?
4. What environmental benefits can the greenery on the roof provide?
5. What are the harmful effects of carbon dioxide in the air on the human?
6. Is heat absorption and insulation a characteristic of Green Buildings?

Part IV Access to Information

Leading Questions

1. Are there standards for Green Buildings?

2. Is natural ventilation a requirement for a building to be considered green?

3. Why does the State promote Green Buildings?

4. Does a higher level of Green Buildings correspond to higher investment costs?

Part V Task Implementation

Leading Questions

1. Can Green Buildings collect rainwater from the roof for reuse?

2. Is the promotion of Green Buildings a collective effort or the responsibility of one person?

3. Does the discharge of indoor living and production wastewater affect the rating of Green Buildings?

4. Which indicators of electricity consumption will affect the rating of Green Buildings?

Part VI Task to Optimization

Leading Questions

1. Please list the Green Buildings you know.

2. Which building materials meet the requirements for Green Building rating?

Module Twenty

Green Building (Ⅱ)

Task 1 Task Description

Task 2 Oral Practice

Task 3 Analysis of Civil Construction Knowledge

Task 1 Task Description

Part I Task Description

Understand the concepts of energy conservation; familiarize yourself with the key aspects of building energy conservation; comprehend the reasons for utilizing green materials; grasp the importance of water conservation; and understand the objectives of resource recycling and utilization.

Part II Learning Goals

1. Master common words in oral English.

2. Understand the common grammar used in oral English.

3. Master the significance of energy conservation and emission reduction.

4. Master the relevant national policies on energy conservation and emission reduction.

5. Understand that energy conservation is also a way to protect nature.

6. Learn ways to save water and develop consciousness of water conservation.

Part III Ideological and Political Points

1. Energy conservation and emission reduction returns the original face of nature.

2. With great love in everyone's heart, the world will be more peaceful.

3. The global shortage of freshwater resources makes water conservation essential for everyone.

4. The use of environmentally friendly plastic pipes is also a way to protect water quality and water resources.

Task 2 Oral Practice

Energy Saving

Kathy—Teacher
Polly—Student
Teddy—Student
Michael—Student
Linda—Student

对话音频

Kathy: Let us continue with the Green Building. Who can review the last part of our knowledge?

Michael: How do the plants on the roof of the building protect the ground environment?

Polly: Clean the air.

Kathy: You are wonderful.

Linda: Continue, please.

Kathy: Energy saving is the main factor of Green Buildings.

Teddy: Where is the energy used in the building?

Kathy: Energy saving means reducing energy consumption. This includes water saving, energy saving, land saving, material saving, and environmental protection.

Linda: What energy will be used in the building?

Kathy: Heating in the winter and refrigeration in the summer.

Polly: Got it. Fuel is needed for heating.

Kathy: That is why we use insulation materials inside or outside the wall.

Linda: How about the roof?

Kathy: We apply insulation materials above the structural slab in a similar way.

Michael: How can we save energy?

Kathy: We can utilize solar energy, air energy, and air sources as heat sources. In terms of electricity consumption, we can use low-energy equipment to save power.

Polly: What types of power-consuming equipment do we have in our school?

Kathy: For example, lamps, air conditioners, fans, and water dispensers.

Teddy: Is there anything else we can do to save energy?

Kathy: We can use green materials, such as plastic pipes. On one hand, they reduce water pollution because they don't rust. On the other hand, their smooth inner walls reduce water head loss in the pipelines, allowing for lower power requirements when pumping water.

Michael: So saving electricity is also saving energy.

Kathy: Up, energy-saving lamps are used widely now.

Linda: Is that all?

Kathy: Of course not. There is much more to cover.

Linda: Can you give us a simplified description?

Kathy: Of course! Would you like a short break?

Polly: No need, please continue.

Kathy: Saving resource consumption is also a way of saving energy.

Linda: You mean reducing cost.

Kathy: Yes. By using low-pollution materials and clean energy. Structural materials must have sufficient strength, durability, and a long-life cycle.

Teddy: Does it mean quality determine price?

Kathy: It is hard to say. A longer service life and less maintenance in the future can reduce the overall cost.

Linda: I understand.

Michael: I see.

Kathy: Water conservation is also an important aspect. We need to save water, considering the possibility of freshwater scarcity in the future.

Polly: It is terrible.

Linda: What do you mean? No water anymore?

Kathy: I will show you water resource and the recycling of water in the world.

Teddy: I am looking forward to it.

Kathy: Going back to saving energy, we should effectively use renewable energy.

Michael: How can we reuse?

Kathy: Such as reusing water through sewage treatment for greening or flushing toilets.

Linda: Wow, it is water conservation.

Kathy: There is still a lot more knowledge to share about saving energy. I will cover it in the future. Let us follow the spirit of the 20th National Congress, which mentions "deepening the energy revolution and accelerating the planning and construction of a new energy system based on China's energy and resource endowment." Class dismissed.

Michael: Thank you!

Task 3 Analysis of Civil Construction Knowledge

Part Ⅰ Related Knowledge Links

Energy conservation and emission reduction	Renewable energy	Green material	Head loss

Part Ⅱ Task Group

<table>
<tr><td>Class</td><td></td><td>Set No.</td><td></td><td>Tutor</td><td></td></tr>
<tr><td>Group Leader</td><td></td><td>Student ID</td><td></td><td></td><td></td></tr>
<tr><td>Team Members</td><td colspan="5"><table>
<tr><td>Name</td><td>Student ID</td><td>Name</td><td>Student ID</td></tr>
<tr><td></td><td></td><td></td><td></td></tr>
<tr><td></td><td></td><td></td><td></td></tr>
</table></td></tr>
<tr><td colspan="6">Division of Tasks</td></tr>
</table>

Part Ⅲ Task Analysis

1. What are the main aspects of building energy conservation?
2. What are the nature energy resources available?
3. How can energy loss be reduced?
4. Can seawater be converted into drinking water?
5. Does pipeline leakage lead to waste of water resources?
6. Can sewage from toilets be directly discharged into rivers?

Part Ⅳ Access to Information

Leading Questions

1. How can energy saving and emission reduction be achieved in daily life?

____

2. What are the methods for water conservation in daily life and work?

3. What are some insulation materials you are familiar with?

4. What is clean energy and how many kinds do you know?

Part V Task Implementation

Leading Questions

1. Can municipal water be used for watering green spaces?

2. Which is more energy-efficient, air energy or solar energy?

3. How can we achieve energy saving in lamp?

4. How can we achieve energy-saving operation of air conditioning systems?

Part VI Task Optimization

Leading Questions

1. Is increasing the thickness of building roofs an effective method for energy saving?

2. How can we achieve energy saving in pump water supply?

Attachment: Evaluation Feedback

Form 1-1 Group Self-Evaluation Score Scale
小组自评打分表

Class 班级		Set No. 组别		Date 日期	
Self-Evaluation Item 自评项目	Self-Evaluation Content 自评内容			Scores 分数	Score Evaluation 分数评定
Information Gathering 知识收集	Effectively use platforms and offline resources to gather useful and relevant information. 有效利用平台、线下资源等收集有用的相关知识			15	
Identification Attitude 认同态度	Be familiar with the workflow and demonstrate an attitude of identification with work. 熟悉工作流程，具有认同工作的态度			15	
Learning Process 学习过程	Proactively learn, maintain good attendance in class, and complete tasks on time. 主动学习，保持良好的上课出勤，并按时完成任务			15	
	Be able to collaborate in learning and think independently, maintaining good information exchange with teachers and classmates. 能够在学习中进行协作和独立思考，与教师和同学保持良好的信息交流			15	
Learning Methods 学习方法	Be able to identify problems, raise issues, analyze problems, and solve them. 能够识别问题，提出议题，分析问题并解决问题			15	
Self-Evaluation Feedback 自评反馈	Be able to speak oral architecture English proficiently and have a good grasp of professional knowledge. 能够流利地表达建筑英语，对专业知识有很好的掌握			25	
Group Self-Rated Scores 小组自评分数					
Experience Sharing 经验分享					
Summary Feedback 总结反馈					

Form 1-2　Teacher evaluation form
教师评价表

Class 班级		Set No. 组别		Name 姓名	
Class Performance 课堂表现					
Evaluation Content 评价内容	Evaluation Item 评价项目		Evaluation Points 评价要点	Scores 分数	Score Evaluation 分数评定
Task Analysis 任务分析	Reading 阅读		Accurate Pronunciation 发音准确	10	
Task Result 任务结果	Answering 答题		Clear Expression 表达清晰	10	
Learning Process 学习过程	Learning Attitude, Attendance 学习态度、出勤情况		Correct Learning Attitude, Good Attendance in Class 学习态度端正、出勤情况良好	15	
Task Implementation 任务实施	Preparation 准备		Various Learning Materials 各种学习资料	10	
	Role Play 角色扮演		Delving into Dialogue Roles 深入对话角色	15	
Task Optimization 任务优化	Group Completion 分组完成		Summary and Analysis 归纳分析	20	
Evaluation Summary 评价总结	Summary 总结		Group Self-Evaluation 小组自评	10	
			Teacher Evaluation 教师评价	10	
Summation 合计				100	

Appendix

模块一　项目决策阶段

单词和短语
Words and Phrases

decision phase [dɪˈsɪʒn] [feɪz] 决策阶段
location [ləʊˈkeɪʃn] 位置
mechanical equipment factory [məˈkænɪkl] [ɪˈkwɪpmənt] [ˈfæktri] 机械设备厂
site selection [saɪt] [sɪˈlekʃn] 选址
crucial [ˈkru:ʃl] 非常重要的
be responsible for [rɪˈspɒnsəbl] 负责
oversee [ˌəʊvəˈsi:] 监督
entire [ɪnˈtaɪə] 整个的
construction process [kənˈstrʌkʃn] [ˈprəʊses] 建设过程
the 20th National Congress agenda [ˈnæʃnəl][ˈkɒŋgres][əˈdʒendə] 第 20 次全国代表大会议程
emphasize [ˈemfəsaɪz] 强调
promotion [prəˈməʊʃn] 提升
carbon reduction [ˈkɑ:bən] [rɪˈdʌkʃn] 降碳
pollution reduction [pəˈlu :ʃn] [rɪˈdʌkʃn] 减少污染
green expansion [gri:n] [ɪkˈspænʃn] 绿色扩张
ecological priority [i:kəˈlɒdʒɪkl] [praɪˈɒrəti] 生态优先
low-carbon [ləʊˈkɑ:bən] 低碳
involved in [ɪnˈvɒlvd] 参与
project proposal [ˈprɒdʒekt] [prəˈpəʊzl] 项目建议书
conduct [kənˈdʌkt] 开展
market research [ˈmɑ:kɪt] [rɪˈsɜ:tʃ] 市场调研
assistance [əˈsɪstəns] 帮助
appreciate [əˈpri:ʃieɪt] 感激
essential [ɪˈsenʃl] 基本的
national policy [ˈnæʃnəl] [ˈpɒləsi] 国家政策
regulation [ˌreguˈleɪʃn] 法规
additionally [əˈdɪʃənəli] 另外
keep up with [ki:p] 跟上

development [dɪˈveləpmənt] 发展
technique [tekˈni:k] 技术
material [məˈtɪərɪəl] 材料
full cooperation [fʊl] [kəʊˌɒpəˈreɪʃn] 全力配合
investment conception [ɪnˈvestmənt] [kənˈsepʃn] 投资构想
Try our best! [traɪ] [ˈaʊə] [best] 我们会尽力！
approve [əˈpru:v] 批准
National Development and Reform Commission [ˈnæʃnəl] [dɪˈveləpmənt] [rɪˈfɔ:m] [kəˈmɪʃn] 国家发展和改革委员会
feasibility study [ˌfi:zəˈbɪləti] [ˈstʌdi] 可行性研究
technical expert [ˈteknɪkl] [ˈekspɜ:t] 技术专家
economic and technological [ˌi:kəˈnɒmɪk] [ˌteknəˈlɒdʒɪkl] 经济和技术的
scientific analysis [ˌsaɪənˈtɪfɪk] [əˈnæləsɪs] 科学分析
demonstration [ˌdemənˈstreɪʃn] 演示
accuracy [ˈækjərəsi] 准确性
furthermore [ˌfɜ:ðəˈmɔ:] 此外
environmental assessment [ɪnˌvaɪrənˈmentl] [əˈsesmənt] 环境评估
safety evaluation [ˈseɪfti] [ɪˌvæljuˈeɪʃn] 安全评价
shorten [ˈʃɔ:tn] 缩短
establish [ɪˈstæblɪʃ] 立项
on precise premise [prɪˈsaɪs] [ˈpremɪs] 在精确前提下
Natural Resources Bureau [ˈnætʃrəl] [rɪˈsɔ:sɪz] [ˈbjʊərəʊ] 自然资源局
strive to [straɪv] 着力于
achieve [əˈtʃi:v] 达到
national objective [ˈnæʃnəl] [əbˈdʒektɪv] 国家目标
energy conservation [ˈenədʒi] [ˌkɒnsəˈveɪʃn] 节能
emission reduction [ɪˈmɪʃn] [rɪˈdʌkʃn] 减排
submit [səbˈmɪt] 递交
geological survey [ˌdʒi:əˈlɒdʒɪkl] [ˈsɜ:veɪ] 地质勘查
promptly [ˈprɒmptli] 立即
preliminary design [prɪˈlɪmɪnəri] [dɪˈzaɪn] 初步设计
separately [ˈseprətli] 分头地

任务 2 译文

模块二　项目设计准备阶段

单词和短语音频

单词和短语
Words and Phrases

design specification [dɪˈzaɪn] [ˌspesɪfɪˈkeɪʃn]	设计标准
hectare [ˈhekteə]	公顷
construction area [kənˈstrʌkʃn] [ˈeəriə]	建筑面积
land use index [lænd] [juːz] [ˈɪndeks]	用地指标
independent workshop [ˌɪndɪˈpendənt] [ˈwɜːkʃɒp]	独立车间
office building [ˈɒfɪs] [ˈbɪldɪŋ]	办公楼
canteen [kænˈtiːn]	食堂、餐厅
staff dormitory [stɑːf] [ˈdɔːmətri]	员工宿舍
detailed description [ˈdiːteɪld] [dɪˈskrɪpʃn]	详细说明
cost saving [kɒst] [ˈseɪvɪŋ]	节约造价
general layout [ˈdʒenrəl] [ˈleɪaʊt]	总图
design profundity [dɪˈzaɪn] [prəˈfʌndəti]	设计深度
take charge of [teɪk] [tʃɑːdʒ]	负责
contact [ˈkɒntækt]	联系
the State Grid [steɪt] [grɪd]	国家电网
power load [ˈpaʊə] [ləʊd]	用电负荷
the Water Company [ˈwɔːtə] [ˈkʌmpəni]	水务公司
the Heating Company [ˈhiːtɪŋ] [ˈkʌmpəni]	热力公司
geological investigation report [ˌdʒiːəˈlɒdʒɪkl] [ɪnˌvestɪˈgeɪʃn] [rɪˈpɔːt]	地勘报告
cooperate [kəʊˈɒpəreɪt]	配合
intensively [ɪnˈtensɪvli]	集中地
people-oriented [ˈpiːpl ˈɔːrientɪd]	以人为本
be responsible for [rɪˈspɒnsəbl]	负责
Green Building [griːn] [ˈbɪldɪŋ]	绿色建筑
energy saving [ˈenədʒi] [ˈseɪvɪŋ]	节能
cost analysis and control [kɒst] [əˈnæləsɪs] [kənˈtrəʊl]	成本分析和控制
sponge city [spʌndʒ] [ˈsɪti]	海绵城市
strive [straɪv]	力争
goal [gəʊl]	目标

deepen [ˈdi:pən] 加深
environmental [ɪnˌvaɪrənˈmentl] 生态环境的
continuously [kənˈtɪnjuəsli] 连续不断地
defense [dɪˈfens] 防护
preliminary design [prɪˈlɪmɪnəri] [dɪˈzaɪn] 初步设计
finalize [ˈfaɪnəlaɪz] 敲定、商定
be prepared separately [prɪˈpeəd] [ˈseprətli] 分头准备

任务 2 译文

模块三　项目初步设计阶段（一）

单词和短语音频

单词和短语
Words and Phrases

scheme designer [ski:m] [dɪˈzaɪnə] 方案设计人
HVAC（Heating, Ventilation and Air Conditioning）
[ˈhi:tɪŋ] [ventɪˈleɪʃn][eə] [kənˈdɪʃənɪŋ] 供暖与通风空调
concept design [ˈkɒnsept] [dɪˈzaɪn] 概念设计
based on [beɪst] 基于
option [ˈɒpʃn] 选择
compare [kəmˈpeə] 对比
rough [rʌf] 粗略的
adhere to [ədˈhɪə] 坚持
the Communist Party of China [ˈkɒmjənɪst] 中国共产党
prosperity [prɒˈsperəti] 繁荣
cultural undertaking [ˈkʌltʃərəl] [ˌʌndəˈteɪkɪŋ] 文化事业
people-centered [pi:plˈsentəd] 以人为本
creative orientation [kriˈeɪtɪv] [ˌɔ:riənˈteɪʃn] 创造导向
excellent [ˈeksələnt] 杰出的
enhance [ɪnˈhɑ:ns] 增强
cost analysis [kɒst] [əˈnæləsɪs] 成本分析
cost estimation [kɒst] [ˌestɪˈmeɪʃn] 成本估算
It' s challenging! [ˈtʃælɪndʒɪŋ] 挑战来了！
spirit of craftsmanship [ˈspɪrɪt] [ˈkrɑ:ftsmənʃɪp] 工匠精神
constantly strive for perfection [ˈkɒnstəntli] [straɪv] [pəˈfekʃn] 精益求精

任务 2 译文

模块四　项目初步设计阶段（二）

单词和短语音频

单词和短语
Words and Phrases

modification [ˌmɒdɪfɪˈkeɪʃn]	修改
the Planning Bureau [ˈplænɪŋ] [ˈbjʊərəʊ]	规划局
feedback [ˈfi:dbæk]	反馈
design review [dɪˈzaɪn] [rɪˈvju:]	设计审查
internally [ɪnˈtɜ:nəli]	内部地
elevation [ˌelɪˈveɪʃn]	立面
section [ˈsekʃn]	剖面
pile position [paɪl] [pəˈzɪʃn]	桩位
foundation [faʊnˈdeɪʃn]	基础
diagram node [ˈdaɪəgræm] [nəʊd]	节点图
power supply [ˈpaʊə] [səˈplaɪ]	供电
transformer substation and distribution station [trænsˈfɔ:mə] [ˈsʌbsteɪʃn] [ˌdɪstrɪˈbju:ʃn] [ˈsteɪʃn]	变配电站
system diagram [ˈsɪstəm] [ˈdaɪəgræm]	系统图
lightning protection [ˈlaɪtnɪŋ] [prəˈtekʃn]	避雷
weak current [wi:k] [ˈkʌrənt]	弱电
air conditioning and ventilation [eə] [kənˈdɪʃənɪŋ] [ˌventɪˈleɪʃn]	空调通风
primary [ˈpraɪməri]	主要的，最初的
heat exchange station [hi:t] [ɪksˈtʃeɪndʒ] [ˈsteɪʃn]	换热站
water supply [ˈwɔ:tə] [səˈplaɪ]	给水
drainage [ˈdreɪnɪdʒ]	排水
violate [ˈvaɪəleɪt]	违反
mandatory provision [ˈmændətəri] [prəˈvɪʒn]	强制性条文
principle [ˈprɪnsəpl]	原则
law-based [lɔ:beɪst]	基于法律的
governance [ˈgʌvənəns]	管理
socialist [ˈsəʊʃəlɪst]	社会主义者
characteristic [ˌkærəktəˈrɪstɪk]	特征
concentrate on [ˈkɒnsntreɪt]	集中精力于
work overtime [wɜ:k] [ˈəʊvətaɪm]	加班

任务 2 译文

模块五　项目施工图设计阶段（一）

单词和短语音频

单词和短语
Words and Phrases

align [əˈlaɪn] 使一致
implement [ˈɪmplɪment] 实施
carbon peak [ˈkɑ:bən] [pi:k] 碳达峰
set up [set] [ʌp] 设置
condition drawing [kənˈdɪʃn] [ˈdrɔ:ɪŋ] 条件图
beam [bi:m] 梁
foundation [faʊnˈdeɪʃn] 基础
the floor plan [flɔ:] [plæn] 楼层平面图
one by one 逐个地
municipal interface [mju:ˈnɪsɪpl] [ˈɪntəfeɪs] 市政接口
relevant department [ˈreləvənt] [dɪˈpɑ:tmənt] 相关部门
reserved hole [rɪˈzɜ:vd] [həʊl] 预留洞
load [ləʊd] 荷载
equipment load [ɪˈkwɪpmənt] [ləʊd] 用电负荷
working progress [ˈwɜ:kɪŋ] [ˈprəʊgres] 工作进度

任务 2 译文

模块六　项目施工图设计阶段（二）

单词和短语音频

单词和短语
Words and Phrases

stringent [ˈstrɪndʒənt]	严格的
comprehend [ˌkɒmprɪˈhend]	理解
embody [ɪmˈbɒdi]	体现
prefabricate [ˌpriːˈfæbrɪkeɪt]	预制
promote [prəˈməʊt]	促进
accelerate [əkˈseləreɪt]	加速
manufacture [ˌmænjuˈfæktʃə]	制造
dormitory [ˈdɔːmətri]	宿舍
complicate [ˈkɒmplɪkeɪt]	复杂化
a piece of cake [piːs] [keɪk]	小菜一碟
coating [ˈkəʊtɪŋ]	涂料
stone paint [stəʊn] [peɪnt]	真石漆
steel structure [stiːl] [ˈstrʌktʃə]	钢结构
internal drainage [ɪnˈtɜːnl] [ˈdreɪnɪdʒ]	内排水
gutter [ˈgʌtə]	天沟
parameter [pəˈræmɪtə]	参数
manufacturer [ˌmænjuˈfæktʃərə]	生产商
central air conditioning [ˈsentrəl] [eə] [kənˈdɪʃənɪŋ]	中央空调
effective height [ɪˈfektɪv] [haɪt]	净高
ceiling [ˈsiːlɪŋ]	吊顶
solar panel [ˈsəʊlə] [ˈpænl]	太阳能板
place [pleɪs]	放置
shower facility [ˈʃaʊə] [fəˈslɪəti]	洗浴设施

任务 2 译文

模块七　项目施工图设计阶段（三）

单词和短语音频

单词和短语
Words and Phrases

pile foundation [paɪl] [faʊnˈdeɪʃn] 桩基
waterproof board [ˈwɔ:təpru:f] [bɔ:d] 防水板
crane beam [kreɪn] [bi:m] 吊车梁
integral assembly type [ˈɪntɪgrəl] [əˈsembli] [taɪp] 整体装配式
truss [trʌs] 桁架
reinforce [ˌri:ɪnˈfɔ:s] 加固
composite slab [ˈkɒmpəzɪt] [slæb] 叠合板
cast–in–place [kɑ:st] [pleɪs] 现浇
in–situ [ɪn ˈsɪtju:] 现场
shear wall [ʃɪə] [wɔ:l] 剪力墙
component [kəmˈpəʊnənt] 构件
seismic fortification intensity [ˈsaɪzmɪk] [ˌfɔ:tɪfɪˈkeɪʃn] [ɪnˈtensəti] 抗震设防烈度
comprehensive [ˌkɒmprɪˈhensɪv] 综合的
security [sɪˈkjʊərəti] 安全
rigorous [ˈrɪgərəs] 严格的
the amount of steel [əˈmaʊnt] [sti:l] 用钢量
frame and shear structure [freɪm] [ʃɪə] [ˈstrʌktʃə] 框剪结构
staircase [ˈsteəkeɪs] 楼梯
precast [ˌpri:ˈkɑ:st] 预制
platform [ˈplætfɔ:m] 平台
high strength steel bar [haɪ] [streŋθ] [ˌri:ɪnˈfɔ:sɪŋ] [sti:l] [bɑ:] 高强度钢筋

任务 2 译文

模块八　项目施工图设计阶段（四）

单词和短语音频

单词和短语
Words and Phrases

fire water consumption [ˈfaɪə] [ˈwɔ:tə] [kənˈsʌmpʃn] 消防用水量
hydrant [ˈhaɪdrənt] 消火栓
sprinkler [ˈsprɪŋklə] 消防喷洒
warehouse [ˈweəhaʊs] 库房
facility [fəˈsɪləti] 设施
raw material [rɔ:] [məˈtɪəriəl] 原料
wooden packing [ˈwʊdn] [ˈpækɪŋ] 木质包装
danger level [ˈdeɪndʒə] [ˈlevl] 危险等级
spraying intensity [ˈspreɪɪŋ] [ɪnˈtensəti] 喷水强度
action area [ˈækʃn] [ˈeəriə] 作用面积
fire flow [ˈfaɪə] [fləʊ] 消防流量
recalculate [ri:ˈkælkjuleɪt] 重新计算
volume [ˈvɒlju:m] 体积
fire reservoir [ˈfaɪə] [ˈrezəvwɑ:] 消防水池
fire pump room [ˈfaɪə] [pʌmp] [ru:m] 消防泵房
municipal water pressure [mju:ˈnɪsɪpl] [ˈwɔ:tə] [ˈpreʃə] 市政水压
Early Suppression Rapid Response [ˈɜ:li] [səˈpreʃn] [ˈræpɪd] [rɪˈspɒns] 早期抑制快速反应
event [ɪˈvent] 事件
political [pəˈlɪtɪkl] 政治的
priority [praɪˈɒrəti] 优先
siphon rainwater [ˈsaɪfn] [ˈreɪnwɔ:tə] 虹吸雨水
drainage [ˈdreɪnɪdʒ] 排水

任务 2 译文

模块九　项目施工图设计阶段（五）

单词和短语音频

单词和短语
Words and Phrases

central air conditioning system [ˈsentrəl] [eə] [kənˈdɪʃənɪŋ] [ˈsɪstəm] 中央空调系统
ventilation [ˌventɪˈleɪʃn] 通风
constant temperature [ˈkɒnstənt] [ˈtemprətʃə] 恒温
humidity [hjuːˈmɪdəti] 湿度
fan [fæn] 风扇
multi-split [ˈmʌlti splɪt] 多联机
pay attention to [peɪ] [əˈtenʃn] 注意
WeChat group [wi:tʃæt] [gru:p] 微信群
innovate [ˈɪnəveɪt] 创新
encourage [ɪnˈkʌrɪdʒ] 鼓励
strengthen [ˈstreŋθn] 加强
dominant [ˈdɒmɪnənt] 显性的
enterprise [ˈentəpraɪz] 企业
innovation [ˌɪnəˈveɪʃn] 创新
exert [ɪgˈzɜ:t] 施加
technology-based [tekˈnɒlədʒi beɪst] 基于技术
backbone [ˈbækbəʊn] 骨干

任务 2 译文

模块十　项目施工图设计阶段（六）

单词和短语音频

单词和短语
Words and Phrases

power consumption index [ˈpaʊə] [kənˈsʌmpʃn] [ˈɪndeks]　用电指标
introduction of two 10kV power supplies [ˌɪntrəˈdʌkʃn] [ˈpaʊə] [səˈplaɪz]　双路供电
cable tray [ˈkeɪbl] [treɪ]　电缆桥架
transformer and distribution room [trænsˈfɔ:mə] [ˌdɪstrɪˈbju:ʃn] [ru:m]　变配电室
municipal [mju:ˈnɪsɪpl]　市政的
overhead [ˌəʊvəˈhed]　架空
bury [ˈberi]　埋地
respect [rɪˈspekt]　尊重
conform [kənˈfɔ:m]　遵守
modernize [ˈmɒdənaɪz]　使现代化
underground floor [ˌʌndəˈgraʊnd] [flɔ:]　地下一层
sufficient [səˈfɪʃnt]　足够的
the State Grid [steɪt] [grɪd]　国家电网
applicable code [əˈplɪkəbl] [kəʊd]　适用规范
the Electric Power Design Institute [ɪˈlektrɪk] [ˈpaʊə] [dɪˈzaɪn] [ˈɪnstɪtju:t]　电力设计院
the design condition [dɪˈzaɪn] [kənˈdɪʃn]　设计条件
the floor height [flɔ:] [haɪt]　层高
on the premise of [ˈpremɪs]　以……为前提
municipal interface [mju:ˈnɪsɪpl] [ˈɪntəfeɪs]　市政接口

任务 2 译文

模块十一 项目施工图阶段（一）

单词和短语音频

单词和短语
Words and Phrases

bid [bɪd] 招标
print [prɪnt] 打印
deliver [dɪˈlɪvə] 发送
internal discussion about the drawings [ɪnˈtɜ:nl] [dɪˈskʌʃn] [ˈdrɔ:ɪŋz] 内部图纸会审
outcome [ˈaʊtkʌm] 结果
open bidding [ˈəʊpən] [ˈbɪdɪŋ] 公开招标
invited tendering [ɪnˈvaɪtɪd] [ˈtendərɪŋ] 邀请招标
announcement [əˈnaʊnsmənt] 公示
qualified [ˈkwɒlɪfaɪd] 合格的
go ahead [gəʊ] [əˈhed] 进行
proceed with [prəˈsi:d] 继续
the Tendering and Bidding Law of the People's Republic of China [ˈtendərɪŋ][ˈbɪdɪŋ] [lɔ:] [ˈpi:plz] [rɪˈpʌblɪk] [ˈtʃaɪnə] 中华人民共和国招标投标法
fairness [ˈfeənəs] 公平
justice [ˈdʒʌstɪs] 公正

任务 2 译文

模块十二　项目施工图阶段（二）

单词和短语音频

单词和短语
Words and Phrases

doubt [daʊt]　疑问

strength [streŋθ]　实力

registration information [ˌredʒɪˈstreɪʃn] [ˌɪnfəˈmeɪʃn]　注册信息

put in great efforts [ˈefəts]　竭尽全力

competitor [kəmˈpetɪtə]　对手，竞争者

extra cautious [ˈekstrə] [ˈkɔːʃəs]　格外谨慎

application for pre-qualification [ˌæplɪˈkeɪʃn] [ˈpriːˌkwɒlɪfɪˈkeɪʃn]　资格预审申请书

rank [ræŋk]　排序

notice of pre-qualification [ˈnəʊtɪs] [ˈpriːˌkwɒlɪfɪˈkeɪʃn]　资格预审合格通知书

survey the site [ˈsɜːveɪ] [saɪt]　勘察现场

send out [send] [aʊt]　发出

pre-tender period [ˈpriːˈtendə][ˈpɪəriəd]　回标时间

receipt [rɪˈsiːt]　回执

optimize [ˈɒptɪmaɪz]　优化

relevant information [ˈreləvənt] [ˌɪnfəˈmeɪʃn]　相关信息

as you instructed [ɪnˈstrʌktɪd]　根据你的指示

answer questions [ˈɑːnsə] [ˈkwestʃənz]　答疑

face to face [feɪs]　面对面

written question [ˈrɪtn] [ˈkwestʃən]　书面问题

clarify and modify [ˈklærəfaɪ] [ˈmɒdɪfaɪ]　澄清和修改

bid security [bɪd] [sɪˈkjʊərəti]　投标保证金

write down [raɪt] [daʊn]　记录

overdue [ˌəʊvəˈdjuː]　逾期

on time [taim]　按时

the list of materials [lɪst] [məˈtɪərɪəlz]　材料清单

the pre-tender estimated price [ˈpriː ˈtendə] [ˈestɪmeɪtɪd] [praɪs]　标底

bid opening [bɪd] [ˈəʊpənɪŋ]　开标

evaluation expert [ɪˌvæljuˈeɪʃn] [ˈekspɜːt]　评审专家

fairness [ˈfeənəs] 公平
winning bidder [ˈwɪnɪŋ] [ˈbɪdə] 中标人
public announcement [ˈpʌblɪk] [əˈnaʊnsmənt] 公示
contract agreement [ˈkɒntrækt] [əˈgri:mənt] 合同协议
interconnect [ˌɪntəkəˈnekt] 相联系
interdependent [ˌɪntədɪˈpendənt] 互相依存的

任务 2 译文

模块十三　项目施工图阶段（三）

单词和短语音频

单词和短语
Words and Phrases

familiarize with [fə'mɪliəraɪz] [wɪð]　熟悉
roughly ['rʌf li]　大致地
make notes [meɪk] [nəʊts]　做备注
feedback ['fi:dbæk]　反馈
technical disclosure ['teknɪkl] [dɪs'kləʊʒə]　技术交底
utility [ju:'tɪləti]　公用的
the Thermal Company ['θɜ:ml] ['kʌmpəni]　热力公司
situation [ˌsɪtʃu'eɪʃn]　情况
asset ['æset]　资产
aspect ['æspekt]　方面
region ['ri:dʒən]　地区
technical experience ['teknɪkl] [ɪk'spɪəriəns]　技术经验
three supplies and one leveling [θri:] [sə'plaɪz] ['levlɪŋ]　三通一平
site lofting [saɪt] ['lɒftɪŋ]　工地放样
supervising engineer ['su:pəvaɪzɪŋ] [ˌendʒɪ'nɪə]　监理工程师
temporary facility ['temprəri] [fə'sɪləti]　临时设施
public facility ['pʌblɪk] [fə'sɪləti]　公用设施
implement ['ɪmplɪment]　落实
labor team ['leɪbə] [ti:m]　劳动班组
deploy [dɪ'plɔɪ]　调配
assignment [ə'saɪnmənt]　任务书
guarantee [ˌgærən'ti:]　保证

任务 2 译文

模块十四　项目施工图阶段（四）

单词和短语音频

单词和短语
Words and Phrases

construction schedule [kənˈstrʌkʃn] [ˈʃedjuːl] 施工进度计划
implement [ˈɪmplɪment] 实施
comprehensive [ˌkɒmprɪˈhensɪv] 综合的
strategy [ˈstrætədʒi] 战略
intensive [ɪnˈtensɪv] 密集的
utilization [ˌjuːtəlaɪˈzeɪʃn] 利用
relevant [ˈreləvənt] 相关的
installation [ˌɪnstəˈleɪʃn] 安装
construction time-frame [kənˈstrʌkʃn] [ˈtaɪm freɪm] 施工时限
tight [taɪt] 紧迫的
potential [pəˈtenʃl] 潜在的
climatic factor [klaɪˈmætɪk] [ˈfæktə] 气候因素
consideration [kənˌsɪdəˈreɪʃn] 考虑
particularly [pəˈtɪkjələli] 特别
duration management [djuˈreɪʃn] [ˈmænɪdʒmənt] 工期管理
milestone node [ˈmaɪlstəʊn] [nəʊd] 里程碑节点
level 1 network plan [ˈlevl] [ˈnetwɜːk] [plæn] 一级网络计划
Gantt chart [ˈgænt] [tʃɑːt] 横道图
unforeseen circumstance [ˌʌnfɔːˈsiːn] [ˈsɜːkəmstəns] 不可抗力因素
extension [ɪkˈstenʃn] 延长
justifiable [ˌdʒʌstɪˈfaɪəbl] 正当的
convincingly [kənˈvɪnsɪŋli] 令人信服地
shortest time-frame [ˈʃɔːtɪst] [ˈtaɪm freɪm] 最短时间
separately [ˈseprətli] 分头

任务 2 译文

模块十五 项目施工图阶段（五）

单词和短语音频

单词和短语
Words and Phrases

position [pəˈzɪʃn]	定位
set out of [set]	开始
disclosure [dɪsˈkləʊʒə]	公开
subcontractor [ˈsʌbkəntræktə]	分包商
woodworking formwork [ˈwʊdwɜːkɪŋ] [ˈfɔːmwɜːk]	木工模板
electric welding [ɪˈlektrɪk] [ˈweldɪŋ]	电焊
reinforcement of column [ˌriːɪnˈfɔːsmənt] [ˈkɒləm]	柱筋
foundation platform [faʊnˈdeɪʃn] [ˈplætfɔːm]	基础承台
people first and prioritizing safety [ˈpiːpl] [fɜːst] [praɪˈɒrətaɪzɪŋ] [ˈseɪfti]	人民至上生命至上
embedding [ɪmˈbedɪŋ]	预埋
accurate [ˈækjərət]	准确的
rework [ˌriːˈwɜːk]	返工
pour [ˈpɔːr]	浇筑
thoroughly [ˈθʌrəli]	彻底
acceptance [əkˈseptəns]	验收
volume of backfill earthwork [ˈvɒljuːm] [ˈbækfɪl] [ˈɜːθwɜːk]	回填土方量
purchase [ˈpɜːtʃəs]	购买
proceed [prəˈsiːd]	继续

任务 2 译文

模块十六　项目施工图阶段（六）

单词和短语
Words and Phrases

construction rectification [kənˈstrʌkʃn] [ˌrektɪfɪˈkeɪʃn]	施工整改
inspection [ɪnˈspekʃn]	查看
crack [kræk]	开裂
pouring plate [ˈpɔ:rɪŋ] [pleɪt]	浇筑板
parallel [ˈpærəlel]	平行的
trench [trentʃ]	地沟
at the edge [edʒ]	在边缘
improper treatment [ɪmˈprɒpə] [ˈtri:tmənt]	处理不当
auxiliary [ɔ:gˈzɪliəri]	辅助的
ditch [dɪtʃ]	沟
analyze [ˈænəlaɪz]	分析
preliminary analysis [prɪˈlɪmɪnəri] [əˈnæləsɪs]	初步分析
constrain [kənˈstreɪn]	约束
shrinkage [ˈʃrɪŋkɪdʒ]	收缩
performance [pəˈfɔ:məns]	性能
prolonged bleeding [prəˈlɒŋd] [ˈbli:dɪŋ]	泌水延长
inadequate [ɪnˈædɪkwət]	不足的
condensation [ˌkɒndenˈseɪʃn]	凝结
environmental [ɪnˌvaɪrənˈmentl]	环境的
solve [sɒlv]	解决
response [rɪˈspɒns]	答复
in written form [ˈrɪtn] [fɔ:m]	以文字形式
initial [ɪˈnɪʃl]	前期的
approach [əˈprəʊtʃ]	方案
rectify [ˈrektɪfaɪ]	整改
impact [ˈɪmpækt]	影响
speed up [spi:d]	加速
nothing more [mɔ:]	没有了

scientifically [ˌsaɪənˈtɪfɪkli] 按科学方法
potential [pəˈtenʃl] 自然灾害
proactively [ˌprəʊˈæktɪvli] 主动地
advance [ədˈvɑːns] 推进

任务 2 译文

模块十七　项目施工图阶段（七）

单词和短语音频

单词和短语
Words and Phrases

crisis [ˈkraɪsɪs]	危险
it is a prophecy [ˈprɒfəsi]	一语成谶
severe [sɪˈvɪə]	严重的
leakage [ˈliːkɪdʒ]	漏失
conduct [kənˈdʌkt]	实施
field maintenance [fiːld] [ˈmeɪntənəns]	现场维护
radiator [ˈreɪdieɪtə]	散热器
get straight to the point [streɪt] [pɔɪnt]	直接说重点
well-installed [wel ɪnˈstɔːld]	安装得很好
be there or be square [skweə]	不见不散
cool down [kuːl]	冷静
such a mess [mes]	一团糟
a 100-year plan and a focus on quality	百年大计质量第一
honestly [ˈɒnɪstli]	坦白地
nonconforming product [ˈnɒnkənˈfɔːmɪŋ] [ˈprɒdʌkt]	不合格产品
overpressure [ˌəʊvəˈpreʃə]	超压
domestic [dəˈmestɪk]	生活的
the devil is in the details [ˈdevl] [ˈdiːteɪlz]	细节决定成败
prioritize [praɪˈɒrətaɪz]	优先处理
assurance [əˈʃʊərəns]	保证
achieve [əˈtʃiːv]	达到
capable [ˈkeɪpəbl]	有能力的

任务 2 译文

模块十八　项目竣工图阶段

单词和短语音频

单词和短语
Words and Phrases

completion report [kəmˈpliːʃn] [rɪˈpɔːt] 竣工报告
construction permit [kənˈstrʌkʃn] [pəˈmɪt] 施工许可证
drawing review comment [ˈdrɔːɪŋ] [rɪˈvjuː] [ˈkɒment] 施工图审查意见
general situation of the project [ˈdʒenrəl] [ˌsɪtʃuˈeɪʃn] [ˈprɒdʒekt] 工程概况
extract [ˈekstrækt] 提取
construction drawing description [kənˈstrʌkʃn] [ˈdrɔːɪŋ] [dɪˈskrɪpʃn] 施工图说明
entire project implementation process [ɪnˈtaɪə] [ˈprɒdʒekt] [ˌɪmplɪmenˈteɪʃən] [ˈprəʊses] 整个项目实施过程
quality acceptance report [ˈkwɒləti] [əkˈseptəns] [rɪˈpɔːt] 质量验收报告
signature [ˈsɪgnətʃə] 签字
seal [siːl] 盖章
appropriate [əˈprəʊpriət] 适当的
stamp [stæmp] 盖章
subgrade [ˈsʌbˌgreɪd] 地基
main structure [meɪn] [ˈstrʌktʃə] 主体结构
meet [miːt] 满足
mandatory standard [ˈmændətəri] [ˈstændəd] 强制性标准
conform to [kənˈfɔːm] 符合
code [kəʊd] 规范
leakage [ˈliːkɪdʒ] 漏失
inspection [ɪnˈspekʃn] 检查
fire protection [ˈfaɪə] [prəˈtekʃn] 消防
acceptance report [əkˈseptəns] [rɪˈpɔːt] 验收报告
retrieve [rɪˈtriːv] 取回
self-evaluation [self] [ɪˌvæljuˈeɪʃn] 自我评估
qualified [ˈkwɒlɪfaɪd] 合格的
Supervision Company [ˌsuːpəˈvɪʒn] [ˈkʌmpəni] 监理公司
comprehensive acceptance [ˌkɒmprɪˈhensɪv] [əkˈseptəns] 综合验收

guided by [ˈɡaɪdɪd] 由……来指导
core [kɔː] 核心
the core socialist values [ˈsəʊʃəlɪst] [ˈvæljuːz] 社会主义核心价值观
culture [ˈkʌltʃə] 文化

任务 2 译文

模块十九　绿色建筑（一）

单词和短语音频

单词和短语
Words and Phrases

Green Building [gri:n] [ˈbɪldɪŋ] 绿色建筑
exterior wall [ɪkˈstɪəriə] [wɔ:l] 外墙
oxygenated [ˈɒksɪdʒəneɪtɪd] 有氧的
no clue [klu:] 完全不知道
complacently [ˈkəmˈpleɪsntli] 自信地
step by step [step] 一步一步地
key point [ki:] [pɔɪnt] 关键点
symbolize [ˈsɪmbəlaɪz] 象征
modern significance [ˈmɒdn] [sɪgˈnɪfɪkəns] 现代意义
Chinese culture [ˌtʃaɪˈni:z] [ˈkʌltʃə] 中国文化
stand for [stænd] 代表
ecology [iˈkɒlədʒi] 生态
environmental protection [ɪnˌvaɪrənˈmentl] [prəˈtekʃn] 环境保护
Lucid waters and lush mountains are invaluable assets [ˈlu:sɪd] [ˈwɔ:təz] [lʌʃ] [ˈmaʊntɪnz] [ɪnˈvæljuəbl] [ˈæsets] 绿水青山就是金山银山
environmental pollution [ɪnˌvaɪrənˈmentl] [pəˈlu:ʃn] 环境污染
defense [dɪˈfens] 防守
eliminate [ɪˈlɪmɪneɪt] 消除
urban [ˈɜ:bən] 城市的
odorous [ˈəʊdərəs] 有气味的
relationship [rɪˈleɪʃnʃɪp] 关系
hold on [həʊld] 少安毋躁
with a smile [smaɪl] 微笑着
take your time 别急
as you all know [nəʊ] 众所周知
purify [ˈpjʊərɪfaɪ] 净化
release oxygen [rɪˈli:s] [ˈɒksɪdʒən] 释放氧气
absorb [əbˈzɔ:b] 吸收
carbon dioxide [ˈkɑ:bən] [daɪˈɒksaɪd] 二氧化碳

rooftop garden [ˈruftɒp] [ˈgɑ:dn]	空中花园
purify [ˈpjʊərɪfaɪ]	净化
vertical rain pipe [ˈvɜ:tɪkl] [reɪn] [paɪp]	雨水立管
intercept [ˌɪntəˈsept]	截留
meanwhile [ˈmi:nwaɪl]	同时
insulation [ˌɪnsjuˈleɪʃn]	隔热
take a break [teɪk] [breɪk]	休息一下

任务 2 译文

模块二十　绿色建筑（二）

单词和短语音频

单词和短语
Words and Phrases

review [rɪˈvjuː] 回顾
energy saving [ˈenədʒi] [ˈseɪvɪŋ] 节能
factor [ˈfæktə] 因素
energy consumption [ˈenədʒi] [kənˈsʌmpʃn] 能量消耗
refrigeration [rɪˌfrɪdʒəˈreɪʃn] 制冷
insulation [ˌɪnsjuˈleɪʃn] 隔热
structural slab [ˈstrʌktʃərəl] [slæb] 结构板
utilize [ˈjuːtəlaɪz] 利用
solar energy [ˈsəʊlə] [ˈenədʒi] 太阳能
air source [eə] [sɔːs] 空气源
in terms of 在……方面
low-energy [ləʊ ˈenədʒi] 低能耗
for example [ɪgˈzɑːmpl] 比如
air conditioner [eə] [kənˈdɪʃənə] 空调
fan [fæn] 风机
water dispenser [ˈwɔːtə] [dɪˈspensə] 饮水机
head loss [hed] [lɒs] 水头损失
pump [pʌmp] 抽水
resource consumption [rɪˈsɔːs] [kənˈsʌmpʃn] 资源消耗
clean energy [kliːn] [ˈenədʒi] 清洁能源
intensity [ɪnˈtensəti] 强度
durability [ˌdjʊərəˈbɪləti] 耐用性
life cycle [laɪf] [ˈsaɪkl] 生命周期
service life [ˈsɜːvɪs] [laɪf] 使用寿命
maintenance [ˈmeɪntənəns] 后期维护
overall cost [ˌəʊvərˈɔːl] [kɒst] 综合造价
looking forward to [ˈlʊkɪŋ] [ˈfɔːwəd] 期待
renewable energy [rɪˈnjuːəbl] [ˈenədʒi] 可再生能源
reuse [ˌriːˈjuːz] 二次回用

sewage [ˈsuːɪdʒ] 污水
green [griːn] 绿化
flushing toilet [ˈflʌʃɪŋ] [ˈtɔɪlət] 冲洗大便器
water conservation [ˈwɔːtə] [ˌkɒnsəˈveɪʃn] 节水
deepen [ˈdiːpən] 加深
revolution [ˌrevəˈluːʃn] 革命
accelerate [əkˈseləreɪt] 加速
resource endowment [rɪˈsɔːs] [ɪnˈdaʊmənt] 能源体系
class dismissed [klɑːs] [dɪsˈmɪst] 下课了

任务 2 译文